Studies in Systems, Decision and Control

Volume 132

Series editor

Janusz Kacprzyk, Polish Academy of Sciences, Warsaw, Poland
e-mail: kacprzyk@ibspan.waw.pl

The series "Studies in Systems, Decision and Control" (SSDC) covers both new developments and advances, as well as the state of the art, in the various areas of broadly perceived systems, decision making and control– quickly, up to date and with a high quality. The intent is to cover the theory, applications, and perspectives on the state of the art and future developments relevant to systems, decision making, control, complex processes and related areas, as embedded in the fields of engineering, computer science, physics, economics, social and life sciences, as well as the paradigms and methodologies behind them. The series contains monographs, textbooks, lecture notes and edited volumes in systems, decision making and control spanning the areas of Cyber-Physical Systems, Autonomous Systems, Sensor Networks, Control Systems, Energy Systems, Automotive Systems, Biological Systems, Vehicular Networking and Connected Vehicles, Aerospace Systems, Automation, Manufacturing, Smart Grids, Nonlinear Systems, Power Systems, Robotics, Social Systems, Economic Systems and other. Of particular value to both the contributors and the readership are the short publication timeframe and the world-wide distribution and exposure which enable both a wide and rapid dissemination of research output.

More information about this series at http://www.springer.com/series/13304

Dipesh H. Shah · Axaykumar Mehta

Discrete-Time Sliding Mode Control for Networked Control System

 Springer

Dipesh H. Shah
Department of Instrumentation
    and Control
Sardar Vallabhbhai Patel
    Institute of Technology
Anand, Gujarat
India

Axaykumar Mehta
Department of Electrical Engineering
Institute of Infrastructure Technology
    Research and Management
Ahmedabad, Gujarat
India

ISSN 2198-4182　　　　　　　ISSN 2198-4190　(electronic)
Studies in Systems, Decision and Control
ISBN 978-981-13-3963-9　　　　ISBN 978-981-10-7536-0　(eBook)
https://doi.org/10.1007/978-981-10-7536-0

Printed on acid-free paper

This Springer imprint is published by the registered company Springer Nature Singapore Pte Ltd.
part of Springer Nature
The registered company address is: 152 Beach Road, #21-01/04 Gateway East, Singapore 189721,
Singapore

To our Parents, Teachers, Family and Friends.

Preface

Networked Control Systems (NCSs) are traditional feedback control loops closed through a real-time communication network. In other words, the exchange of information such as reference input or set point, plant output or sensor data and controller output between control system components (sensors, controllers, actuators) is carried out via a communication network. NCS has become popular in the field of control due to its distinct advantages such as low cost, reduced weight, simple installation and maintenance, resource sharing and high reliability. Moreover, NCS has got wide industrial applications such as in manufacturing plants, smart grid, haptic collaboration, vehicles, aircraft, robotics, spacecraft. NCS generally possesses a dynamic nature, which results in various challenges for researchers such as network-induced time delay, packet loss, packet disordering, resource allocation and bandwidth sharing. It is well known that the performance of NCS is significantly deteriorated due to these communication irregularities if these challenges are not handled properly. Among all these issues, network-induced time delay and packet loss are considered to be crucial issues in NCS that deteriorate the stability and performance of closed-loop control systems significantly.

The network-induced delays may be constant, time-varying and, in most cases, random. The nature of network-induced delay mainly depends on the configuration of the communication medium. If the communication medium is configured using leased lines concept, then the delays are always deterministic in nature. And whenever the communication medium is shared by a large number of devices, then the delays are random in nature. It is worth to mention here that the amount of time required for the data packets to travel from sensor to controller and controller to actuator is defined as total network delay. The controller mainly suffers from sensor to controller delay. When such network-induced delays are transformed into discrete-time domain, it mostly possesses non-integer type of values. Such network-induced delays in discrete-time domain are referred to fractional delays, which may be either deterministic or random in nature. So, it is important to compensate the effect of deterministic as well as random fractional delays in discrete-time domain at each sampling instant.

Further, as mentioned above, there are also possibilities of packet loss/information loss during the transmission of data packets from sensor to controller as well as controller to actuator. The packet loss usually takes place due to heavy network load, network congestion and node competition. In discrete-time domain, the network-induced delay greater than one sampling time is also considered as packet loss. The nature of network-induced delay and single packet loss as well as multiple packet loss is mainly dependent on the configuration of network medium.

In recent years, many algorithms have been studied for the stability analysis and controller design for NCS that include PI controller, state feedback controller, H_∞ controller, model predictive controller, sliding mode controller. Among them, sliding mode controller (SMC) is one of the robust control algorithms due to its invariance properties to parameter variation and uncertainties.

This monograph presents some novel algorithms for designing discrete-time sliding mode controller (DSMC) for NCS having both types of fractional delays, i.e. deterministic and random, along with different packet loss conditions, i.e. single packet loss and multiple packet loss. The efficacy of the proposed control algorithm is tested with real-time networks such as CAN and Switched Ethernet medium and experimentally verified by DC servomotor. The robustness of the proposed discrete-time sliding mode controller is improved through disturbance estimator in the presence of multiple packet transmission policy and matched uncertainty.

This monograph contributes mainly the following:

- In Chap. 1, the introduction of Networked Control System in continuous- and discrete-time domains that include time delay compensation methods and design of controllers with single packet loss and multiple packet loss is briefly discussed.
- In Chap. 2, preliminaries and literature survey of NCS and SMC technique in continuous- and discrete-time domains are presented.
- In Chap. 3 and Chap. 4, a modified discrete-time sliding surface and discrete-time sliding mode controller are proposed using the compensated state information that encompasses deterministic type fractional delay and single packet loss. The proposed algorithms are also compared with conventional sliding mode controller using CAN and Switched Ethernet as network medium.
- In Chap. 5, the multirate output feedback approach for the state estimation in the closed loop is incorporated. The main advantage of using multirate output feedback approach is that the system states are computed based on the output information available and the error between computed and estimated state variables goes to zero exactly after one sampling instant even in the presence of networked delay.
- In Chap. 6, discrete-time sliding surface is designed for random fractional delay and single packet loss that occur within the sampling period. The random delay is compensated using Thiran's approximation technique in the presence of packet loss situation. The efficacy of proposed non-switching type of DSMC is endowed by simulation results and also experimentally validated on servo system.

- Further in Chap. 7, the proposed algorithm is extended for random fractional delay with multiple packet loss situation. The disturbance estimator is designed that estimates the disturbance signal and improves the performance of the system. The efficacy of the algorithm is endowed by the simulations under various fractional delays and matched uncertainties.
- In Chap. 8, concluding remarks, future scope and challenges in NCSs are presented.

Keywords Networked control system, Discrete-time sliding mode control, Fractional delay, Packet loss, Multirate output feedback, Disturbance estimator.

Ahmedabad, India   Dipesh H. Shah
September 2017   Axaykumar Mehta

Contents

About the Authors

Dipesh H. Shah was born in Vadodara, Gujarat, India, in 1985. He completed his B.E. (Instrumentation and Control), M.E. (Applied Instrumentation) and Ph.D. (Networked Control System) in 2007, 2010 and 2018 from Gujarat University and Gujarat Technological University, Ahmedabad respectively. He is currently working as an Assistant Professor in Instrumentation and Control Department at Sardar Vallabhbhai Patel Institute of Technology, Vasad, Gujarat, India. He is also serving as an Faculty Advisor at Institute level for ISA student' chapter. He has published various research papers in peer-reviewed journals and presented several papers in national and international conferences. His research interest includes discrete-time sliding mode control, networked control system and communication protocols.

Axaykumar Mehta was born in Bharuch, a small town in south Gujarat, India, in 1975 and received his B.E. Electrical (1996), M.Tech. (2002) and Ph.D. (2009) degree from Gujarat University Ahmedabad, Indian Institute of Technology Kharagpur and Indian Institute of Technology Bombay respectively. Currently he is an Associate Professor at the Institute of Infrastructure Technology Research and Management, Ahmedabad, Gujarat, India since 2014. Prior to that he has worked at various faculty positions in different colleges of Gujarat for teaching undergraduate and postgraduate courses during 1996–2011. He has acted as Professor and Director of the Gujarat Power Engineering and Research Institute, Mehsana, Gujarat, India, from 2012 to 2014. He was also associated with Indian Institute of Technology Gandhinagar from 2010 to 2011 an Associate Faculty. He is recipient of research grant on "Control and Diagnosis of Multi-Agent Systems" from University of Cagliari, Italy 2015–16. He has supervised 03 (completed) and 02 (on-going) students their Ph.D. and several masters thesis. His major research interests are nonlinear sliding mode control and observers, application of sliding mode control in electrical engineering, Networked Control Systems and control of Multi-Agent System. He has published 50 research papers in peer-reviewed international journals and conferences of repute. He is author for three monographs/books and has published three patents with Indian patent office. He is a Senior Member of IEEE, Life Member of the Institution of Engineers India, Life Member of the Indian Society for Technical Education and Life Member of Systems Society of India (SSI). He is also conferred 2014 Best Pedagogical Innovation award 2014 by Gujarat Technological University, Ahmedabad, India for his contribution in online pedagogy.

Acronyms

Abbreviations

CAN	Controller area network
CSMA/CD	Carrier sense multiple access collision detection
DNSMC	Discrete-time networked sliding mode control
DSMC	Discrete-time sliding mode control
FIFO	First-in, first-out
GSM	Gain schedule middleware
LAN	Local area network
LQG	Linear–quadratic–Gaussian
LQR	Linear–quadratic regulator
LTI	Linear time-invariant
MAN	Metropolitan area network
MROF	Multirate output feedback
NCS	Networked Control System
PAN	Personal area network
RMPC	Robust model predictive control
RTT	Round-trip time
SISO	Single input, single output
SMC	Sliding mode control
TOD	Try-one-discard
VSC	Variable structure control
WAN	Wide area network
WSN	Wireless sensor networks
ZOH	Zero-order hold

Symbols

$x(t)$	Plant state vector in continuous-time domain
$u(t)$	Control input signal in continuous-time domain
$d(t)$	Slow time-varying disturbance signal in continuous-time domain

$y(t)$	Output signal in continuous-time domain
$x(k)$	Plant state vector in discrete-time domain
$u(k)$	Control input signal in discrete-time domain
$d(k)$	Disturbance signal in discrete-time domain
$y(k)$	Output signal in discrete-time domain
A	System matrix in continuous-time domain
B	Control input matrix in continuous-time domain
C	Output matrix
F	System matrix in discrete-time domain
G	Input matrix in discrete-time domain
a_1, a_2, M_s, c	Constants
$s(t)$	Sliding surface in continuous-time domain
$s_d(k)$	Priori function
k_s, ι, ψ	User-defined constant
δ_0	Positive offset
p_s, k'	Positive integer
k_t	User-defined gain
k_t^+, k_t^-	Lower and upper bounds coefficients
$\hat{S}$	Model uncertainty
S_1, S_2	Mean and deviated value of $\hat{S}$
S_u, S_l	Upper and lower bounds of $\hat{S}$
λ_s	Switching gain
τ_t	Total delay which includes system- and network-induced delay in continuous-time domain
τ	Total network-induced delay (feedback and forward channel delay) in continuous-time domain
τ_{sc}	Sensor to controller delay in continuous-time domain
τ_{ca}	Controller to actuator in continuous-time domain
h	Sampling interval
τ'	Deterministic total network-induced fractional delay
τ'_{sc}	Sensor to controller deterministic type fractional delay
τ'_{ca}	Controller to actuator deterministic type fractional delay
τ_r	Total random network delay in continuous-time domain
$\hat{\tau}$	Total random network fractional delay in discrete-time domain
τ_{rsc}	Random sensor to controller delay in continuous-time domain
τ_{rca}	Random controller to actuator delay in continuous-time domain
$\hat{\tau}_{sc}$	Random sensor to controller fractional delay in discrete-time domain
$\hat{\tau}_{ca}$	Random controller to actuator fractional delay in discrete-time domain
$\hat{\tau}_l$	Lower bound of random fractional delay
$\hat{\tau}_u$	Upper bound of random fractional delay
d_l	Lower bound of disturbance
d_u	Upper bound of disturbance

τ_p	Total processing delay in continuous-time domain
τ_{sp}	Sensor processing delay in continuous-time domain
τ_{cp}	Controller computational delay in continuous-time domain
τ_{ap}	Actuator processing delay in continuous-time domain
v	Fractional part of delay
l	Order of approximation
δ	Signal transmission delay
$s(k)$	Sliding variable in discrete-time domain
C_s	Sliding gain
α	Parameter calculated using Thiran's approximation
Q, R	Matrices of appropriate dimensions in LQR
ε, q	User-defined constants of Gao's law
sgn	Signum function
d_1	Mean value of disturbance
d_2	Deviated value of disturbance
d_s	Compensated disturbance signal
$\eta, \beta, \rho, \gamma$	Smallest parameter constant obtained using Lyapunov stability analysis
V_s	Lyapunov function
Φ, κ	Stability parameters
α'	Parameter computed based on actuator to controller fractional delay
u_a	Compensated control signal at actuator side
T_s	Settling time
τ_{CAN}	Amount of network delay in CAN medium
τ_{ETHER}	Amount network delay in Switched Ethernet medium
$\theta(s)$	Output of the system (position)
V_m	Input to the system
J_m	Rotor inertia
R_m	Terminal resistance
K_m	Motor back emf constant
$\bar{\alpha}, \bar{\beta}$	Probability of state and control data packet loss
$x'_c(k)$	Communicated state variable over the network
$\{d_1, d_2, .., d_q\}$	Values in a finite set
β_v	Positive scalar quantity
$\alpha(k), d_v$	Stochastic variables
$E\{d_v\}$	Expectation of stochastic variable d_v
w	Number of trials
λ	Average number of events per interval
e	Euler's constant
ς, γ'	Random parameter generated using Thiran's approximation
$E\{\alpha(k)\}$	Expectation of stochastic variable $\alpha(k)$
$u_c(k)$	Communicated control signal over the network

Γ	Stability parameter computed using packet loss and random fractional delay
τ_c	Sampling rate of control input signal
ζ	Sampling rate of output signal
F_ϕ	System matrix sampled at ϕ sampling interval
G_ϕ	Control input matrix sampled at ϕ sampling interval
Λ	Positive integer
F_ζ	System matrix sampled at ζ sampling interval
G_ζ	Control input matrix sampled at ζ sampling interval
$\hat{x}$	Estimated state variable computed using multirate output feedback approach
y_k	Output stack
$\hat{\tau}_{sc_{max}}$	Max delay experienced by the packet as the sensor to controller delay
$\hat{\tau}_{sc_i}$	Sensor to controller fractional delay generated from i^{th} sensor
$\rho_1..,\rho_n$	Random variables uniformly distributed over the interval [0,1]
$P_{loss1},..,P_{lossn}$	Probability of multiple state packet loss over the network
ζ_{max}	Parameter designed using Thiran's approximation for max delay experienced by the packet
$\hat{d}_s(k)$	Compensated estimated disturbance applied to the network
$\hat{d}(k)$	Output of disturbance estimator

List of Figures

List of Tables

Chapter 1
Introduction

Abstract In this chapter, brief introduction and literature survey for Networked Control Systems (NCSs) are presented. The chapter includes detailed literature survey for NCS that covers control algorithms, compensation methods for network delay with packet loss and mathematical modelling of network delays and packet loss in continuous-time as well as discrete-time domain.

Keywords Networked control system · Sliding mode control
Networked delay · Packet loss

1.1 Introduction

1.1.1 Brief Introduction to Networked Control System

The advent of communication networks introduced the concept of remotely controlling a system which gave birth to Networked Control Systems (NCSs). The classical definition of NCSs [1] can be as follows: when a traditional feedback control system is closed via a communication channel, which may be shared with other nodes outside the control system, then the control system is called a Network Control System (NCS). An NCS can also be defined as a feedback control system wherein the control loops are closed through a real-time network [2, 3]. The defining feature of NCS is that information (reference input, plant output, control input, etc.) is exchanged using a communication network among control system components (sensors, controllers, actuators, etc.). The conceptual model of NCS is shown in Fig. 1.1 [2]. The networked medium can be wired or wireless depending on the type of the applications. In NCS, when any form of data that is transmitted through wires, then such medium is called as wired network, while any form of data that is transmitted without use of electrical conductor, then such medium is called as wireless network medium. The main advantage of using wired medium is data security. However, the main advantage of using the wireless network medium is to get rid of wires. In NCS, the wired communication is carried out through CAN, Switched Ethernet, Ethernet, Profibus and

D. H. Shah and A. Mehta, *Discrete-Time Sliding Mode Control for Networked Control System*, Studies in Systems, Decision and Control 132,
https://doi.org/10.1007/978-981-10-7536-0_1

Fig. 1.1 Conceptual model
of NCS [2]

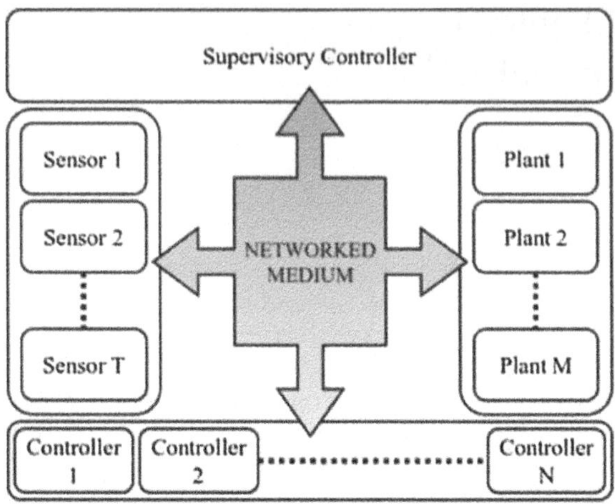

Profinet networked medium, while wireless communication is done through wireless
LAN, wireless PAN, wireless MAN or wireless WAN [3].

1.1.2 Advantages and Applications of Networked Control System

For many years now, data networking technologies have been widely applied in
industrial and military control applications. These applications include manufacturing plants, automobiles and aircraft. Connecting the control system components
in these applications such as sensors, controllers and actuators via a network can
effectively reduce the complexity of systems with nominal economical investments.
Furthermore, network controllers allow data to be shared efficiently without bulk
wiring. It also allows easily to add more sensors, actuators and controllers with
very little cost and without heavy structural changes to the whole system. Most
importantly they connect cyberspace to physical space making task execution from
a distance easy. These systems are becoming more realizable today and have a lot
of potential applications including space explorations, terrestrial exploration, fac-
tory automation, remote diagnostics and troubleshooting, hazardous environments,
experimental facilities, domestic robots, automobiles, aircraft, manufacturing plant
monitoring, nursing homes or hospitals, telerobotics, smart grid.

1.1.3 Structure of Networked Control System

In general, there are two major types of control systems that utilize communication networks: (1) shared network control system and (2) remote control system.
Figure 1.2 shows the architecture of shared network control system [2]. It can be

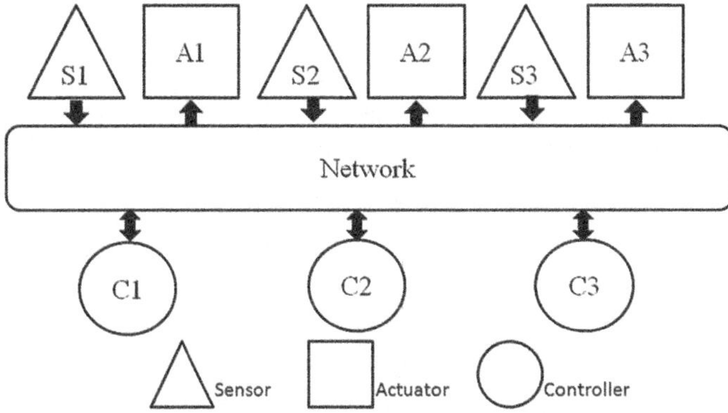

Fig. 1.2 Shared structure of NCS [2]

noticed that with shared network control system, the transfer of information from sensors to controllers and control signals from controllers to actuators greatly reduces the complexity of connections and provides more flexibility in installation, ease of maintenance and troubleshooting. Moreover, it also provides the communication among control loops [2, 4, 5]. This feature is extremely useful when a control loop exchanges information with other control loops to perform more sophisticated controls, such as fault accommodation and control. Similar structures for network-based control have been applied to automobiles and industrial plants. The other control system that utilizes the network medium is remote control system. In remote control system, the place where central controller is installed is called a local site and the place where plant is installed is called a remote site. The data transfer between local site and remote site is carried out through communication network. Sometimes the remote control system is also defined as teleoperation control system. There are two general approaches to design an NCS using remote control system: (i) hierarchical structure and (ii) direct structure. In hierarchical structure, there are several subsystems that are connected to central controller through communication network. Each subsystem contains sensor, actuator and controller by itself as depicted in Fig. 1.3 [2]. In this case, a subsystem controller receives a set point from the central controller. The subsystem then tries to satisfy this set point by itself. The sensor data or status signal is transmitted back via network to the central controller. The block diagram of direct structure is shown in Fig. 1.4 [2]. In this case, the sensor and actuator are attached to a plant, while a controller is separated from the plant by a network connection. The sensor transmits the signal to the controller through the network medium, and controller sends back the processed control signal to the plant via actuator through the network medium. This type of configuration is used mainly in the process industries and haptic surgery. Many complex network control systems use the combination of both the structures known as hybrid structure.

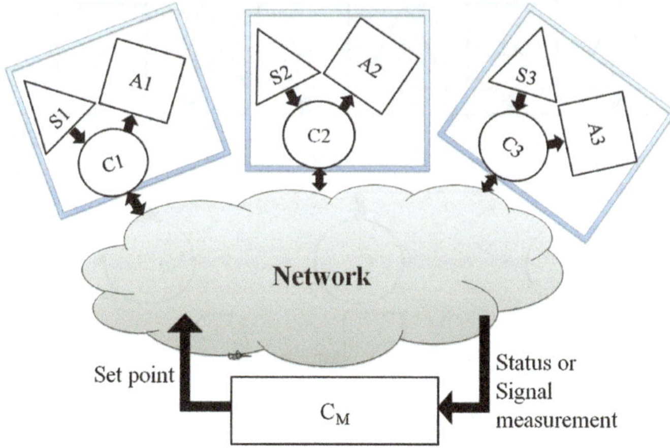

Fig. 1.3 Hierarchical structure of NCS

Fig. 1.4 Direct structure of
NCS

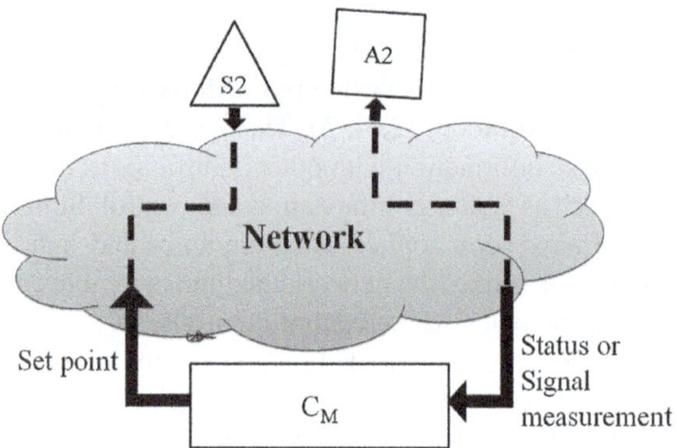

1.1.4 Concerns in Networked Control System

The presence of communication medium in Networked Control System leads to
several natural issues such as:

- **Time delay**: The time required for the data to travel within the network is defined
  as time delay. The nature of time delay depends on the various factors such as
  network configuration, distance of communication between plant and controller,
  baud rate, network characteristics and network topology. The time delay can affect
  the performance of the system in all the structure of NCSs (shared, hierarchical
  and direct).
- **Packet Loss**: Whenever the data transmitted from sensor or controller through the
  network and fails to reach the destination, then such condition is defined as packet

loss condition. The packet loss is mainly caused due to congestion, network traffic and jitter problems. There are two types of packet loss: (i) single packet loss and (ii) multiple packet loss. The packet loss situation occurs in all the structure of NCSs (shared, hierarchical and direct).

- **Packet disorder**: The packet disorder issue is generally caused in wireless NCS due to heavy traffic, congestion or jitter. In wireless NCS, the communications take place in the form of small packets. So in order to have secure communication, each packet is provided with a unique identification number in the header. During transmission, if any packet is lost and fails to reach at the destination, the packet disorder situation takes place. If this disorder is not corrected, then it severely affects the performance of the closed-loop system. This situation also takes place in all the structure of NCSs connected wirelessly.
- **Bandwidth Sharing**: This issue occurs in both the shared structure as well hierarchical structure of NCSs. Both these structures provide the flexibility of connecting large number of devices (such as plant, controller, sensor and actuator) through a common network medium. As the number of devices increases, the bandwidth sharing is also increased which in turn causes reduced transmission speed, congestion, jitter or networked traffic problem. This may further deteriorate the system performance.
- **Security**: The security is one of the major concern in NCS when the communication is carried out without wires. In wireless communication, there are chances of hacking due to which the false data is generated at the controller side and may cause the instability in the system. This issue needs to be handled very appropriately when the communication is carried out through shared structure or hierarchical structure of NCS.

1.2 Literature Review on Networked Control System

Due to its distinct advantages and wide industrial applications, NCS has become popular among the control engineers and also it has become an active research topic among international researchers fraternity. As mentioned earlier, NCS generally possesses a dynamic nature which results in various challenges for researchers in terms of random time delay, packet loss, multiple packet loss, packet disordering, resource allocation and bandwidth sharing. If these challenges are not handled properly, they may result in degradation of the system's performance. Among these challenges, time delay and packet loss are considered to be crucial issues in NCS that causes potential instability.

The next section presents the concise literature survey on the compensation of network-induced delay and packet loss in continuous-time domain as well discrete-time domain.

1.2.1 NCS in Continuous-Time Domain

Various researchers [6–19] have laid their sincere efforts for designing different control algorithms that compensates the effect of network delay. In the early stages of NCS, when modelling of random time delay was difficult to obtain, the most appropriate approach was to treat the random time delay as constant which is called as deterministic delays. Luck and Ray [6, 7] introduced the concept of compensating the time delay in continuous-time domain. They compensated the effect of time delay by introducing the receiver buffer at the controller and actuator side. The size of the buffer was equal to sensor to controller delay and controller to actuator delay. The proposed methodology was tested under IEEE 802.4 network test bed considering the deterministic types of delays. Later on, Luck and Ray [8, 9] also designed predictor-based compensator in which observer was designed to estimate the plant states and predictor was used to predict the control sequences based on the past input signals. The FIFO buffer was set at the controller side and actuator side that stores the past output measurements as well as control measurements. The size of the buffer was set according to the upper bounds of sensor to controller delay and controller to actuator delay. They also tested the efficacy of the proposed algorithm on IEEE 802.4 networked medium. Chen [10] designed conventional form of memory feedback controller based on delay compensation method. Yu et al. [11] designed multiple step delay compensator for NCS in the presence of dynamic noise and measurement noises. Yang [12] proposed the state feedback controller in the presence of network delays in continuous-time domain. They proposed ZOH model at controller and actuator side to compensate the effect of network delay. They also assumed that the sensor is time-driven device, while actuator is event-driven device. Montestruque and Antsaklis [13] designed state feedback controller that compensates the effect of deterministic network delays in continuous-time domain. Kim et al. [15] modelled an NCS as a switched system with constant input delays and derived the sufficient conditions for the system stability using piecewise continuous Lyapunov methods. Godoy et al. [16] designed PID controller to compensate the network delay and validate the feasibility of controller through DC motor as plant and controller area network (CAN) bus as networked medium. Li et al. [17] designed a method for Internet-based network control system in a dual rate configuration to achieve load minimization and dynamic performance specifications. The remote PID controller was designed which regulates the output according to desirable reference and adopts the lower sampling rate to reduce the load on the network. The performance of the system was validated for fixed network delays.

In view of this, an increasing number of researchers began to investigate different control methodologies for NCSs with random or time-varying delays. Vardhan and Kumar [18] used smith predictor algorithm to compensate the effects of time-varying network delays in continuous-time domain. Urban et al. [19] studied the effect of network delays in wired and wireless networked medium using PID controller. They used CAN protocol for the wired communication and Zigbee protocol for wireless networked medium. Zhang et al. [20] proposed the stability criteria for NCS having network delays shorter as well as longer than sampling interval. They also proposed state feedback controller using conventional estimator technique that compensates the effect of network delays having time-varying nature. Similarly, Walsh et al. [21] proposed the mathematical model of NCS considering time-varying network-induced delay. They derived the stability criteria for general NCS in continuous-time domain based on TOD (try-once-discard) algorithm. Yue et al. [14] designed state feedback controller in the presence of time-varying network delays. They assumed that the network delays are lesser than sampling interval. Tipsuwan and Chow [22] proposed the concept of external gain scheduling via GSM. The GSM was used to adjust the controller gains externally at the controller output with respect to the current network traffic conditions without interrupting the internal design of controller. The network delays in the forward channel and feedback channel were modelled using RTT approach. Ji Kim [23] proposed state feedback control with estimator to compensate the effect of time-varying network delay in the presence of matched uncertainty. They tested the efficacy of the proposed controller using Ethernet as a network medium. Ma and Zhao [24] derived the stability criteria for closed-loop NCS using the average dwell time approach and piecewise Lyapunov function method. They designed state feedback controller with estimator that takes care of sensor to controller delay. Peng and Yue [25] designed the state feedback controller for NCS considering time-varying network delay in the states and matched uncertainty. Gao et al. [26] proposed a new time delay system approach which contains multiple successive delay components in the plant states, and based on that, they designed the H_∞ controller to overcome the effect of these state delays. Liu et al. [27] designed network predictive controller to overcome the effects of random network delay in continuous-time domain. The effects of random delays were compensated through network delay compensator placed on the actuator side. The network delay chooses the control input values from the control latest prediction sequence. Cuellar et al. [28] proposed an observer-based predictor using the Pade approximation technique for time lag processes. Sun and Xu [29] modelled the random time delays using stochastic approach in continuous-time domain. They used Markov jump linear systems approach to model sensor to controller random delay, while controller to actuator delay was assumed to be constant. Yuhong and Yeguo [30] designed state feedback controller considering time-varying network delay in the states and proved the closed-loop NCS stability using LMI approach. Ono et al. [31] designed a state feedback controller based on a modified Smith predictor which stabilized the plant in the presence of dead time. Similarly, Ridwan and Trilaksono [32] designed the H_∞ state feedback controller assuming all state variables are measurable in the presence of time-varying network delays. Vallabhan et al. [33] have used the analytical framework approach for compensation of random time delay and packet loss. Hikichi et al. [34] worked on continuous-time delay compensation using predictors and disturbance observer for designing a PID controller. Hu et al. [35] designed a sliding mode intermittent controller for bidirectional associative memory (BAM) using neural networks with delays. Cac et al. [36] used a pole placement method for compensating the time delay in the continuous-time domain. The algorithm was designed for the CAN-type deterministic networked medium. Yi et al. [37]

solved the time delay problem by using the Smith predictor algorithm. The method was verified over wireless sensor networks (WSN) connected between the controller output and plant input. Recently, Khanesar et al. [38] modelled the random time delays using a uniform probability distribution function in continuous-time domain. Saravanakumar et al. [39] proved the stability using a Markovian jump approach for neural networks having varying time interval delays.

1.2.2 NCS in Discrete-Time Domain

Like continuous-time domain, many researchers [40–52] have also tried to focus their work in discrete-time domain. Jacovitti and Scarano [40] proposed various time delay estimation techniques for discrete-time systems. Nilsson et al. [41] used stochastic approach to design state feedback controller for time-varying network delays in discrete-time domain. Similarly, Shousong et al. [42] also used stochastic approach for designing optimal controllers for NCS. They assumed that the random network delays are greater than sampling interval. Zhivoglyadov et al. [43] proposed state feedback observer technique for linear network control system to compensate the effect of random delays. Yue et al. [44] provided the model of NCSs with random network-induced delay in discrete-time domain. They designed H_∞ controller to compensate the effect of random delays in the presence of matched uncertainty. Zhao et al. [45] designed integrated predictive controller for Networked Control System. The predictive controller is applied to generate the control predictions for each delayed sensing data and previous control information. They also designed the time delay compensator at actuator side that actively compensates the forward channel delay when control action is taken. Gou [46] designed the state feedback controller in discrete-time domain based on variable-period sampling approach for random network delays in NCS. Xiong et al. [47] introduced the concept of ZOH model at controller and actuator side that compensates the effect random network delays in discrete-time domain. The proposed ZOH model has an capability of choosing the newest control input. Li et al. [48] designed a sliding mode predictive control for compensation of delay in a Networked Control System using a Kalman predictor. They considered networked delays are random in nature with an integral multiple of sampling interval. Guo et al. [49] considered the state estimation problem for wireless NCS. The sliding mode observer was designed to solve the state estimation problem considering stochastic uncertainty and time delay. Yao et al. [50] designed a robust model predictive control (RMPC) and state observer for a class of time-varying systems under input constraints such as matched uncertainty. Yang et al. [52] proposed discrete-time sliding mode observer that estimates the random delay and compensates its effect in the presence of matched uncertainty. Argha et al. [51] designed stochastic-type sliding mode controller that compensates the effect of random networked delay with values lesser than sampling interval.

Various researchers [41, 43, 46, 51, 53–57] have also laid their sincere efforts to model random time delays in last decade. Among them, Nilsson et al. [41] introduced the time stamp technique to model the random time-varying networked delay. Shousong et al. [42] used stochastic approach to model the random network delays. Gou [46], Zhang et al. [53] as well as Dong and Kim [54] modelled random time delay using the concept of Markov's chain process in discrete-time domain. They have used two-state Markov's chain model to describe sensor to controller delay and controller to actuator delay. While Yang et al. [55] modelled random networked delay using Bernoulli's distributed white sequence approach. Shi and Yu [56] modelled random delays using Markov's chain process and designed output feedback controller to handle the effects of random delay. Ge et al. [57] used an independent and identically distributed approach to model the time-varying networked delay and proposed state feedback controller. Recently, Argha et al. [51] proposed Bernoulli's white sequence approach for modelling the random time delay and proposed sliding mode controller in the presence of random time delay and matched uncertainty.

1.2.3 Packet Losses in NCS

As mentioned above, there are also possibilities of packet loss during the transmission of data packets from sensor to controller as well as controller to actuator. The packet loss takes place due to heavy network load, network congestion and node competition. In NCS, there are two types of packet losses (i) single packet loss and (ii) multiple packet loss. The single packet loss situation generally occurs when the communication of data transfer is done over a shorter distance, and multiple packet loss situations generally occur when communication of data transfer is done over a longer distance. In the research works [20, 38, 44, 48, 51, 53–64], mathematical model is proposed assuming that the packet loss within the communication medium occurs when network delay is greater than sampling interval. Zhang et al. [20] consider the deterministic single packet loss model with packet dropouts occurring at an asymptotic rate. Khanesar et al. [38] derived the packet loss model using the concept of uniform distributed probability function with single packet loss assumption. Yue et al. [44] designed single and multiple packet loss model in context with random network delays. They assumed that whenever the controller and actuator are not updated, the data packet loss takes place for that sampling interval. Li et al. [48] designed sliding mode predictive controller under multiple packet transmission policy. Zhang et al. [53] considered the packet loss in correspondence with random time delay model. They also made a generalized assumption that when the delays are greater than sampling interval, the data packets will be lost at the controller side. Dong and Kim [54] used Dirac delta probability function to derive the mathematical model of packet loss assuming the single packet loss situation. Similarly, Yang et al. [55] also considered the same packet loss approach while modelling the random time delays. Shi and Yu [56] assumed the packet loss situation while modelling the random time delays using Markov's chain process. Argha et al. [51] included the random packet loss situation while modelling the random time delays using Bernoulli's distribution. Hespanha et al. [58] used Bernoulli's probability distribution function to derive the mathematical model of single packet loss as well as multiple packet loss. In both the cases, the situation of packet loss was considered when the network delay is greater than sampling interval. Gupta et al. [59] designed optimal LQG controller that compensates the effect of packet loss occurring within the network. Similarly, Wu and Chen [60] used the concept of the state estimation to compensate the effect of packet loss in discrete-time domain in NCS. Zhu and Yang [61] designed state feedback controller with multiple packet transmission. They proposed the model of NCS with multiple packet transmission and given mathematical model of packet dropout in sensor to controller channel and controller to actuator channel. Niu and Ho [62] designed the compensator using probability function that compensates the effect of packet loss within the network. Wen and Gao [63] proposed H_∞ controller for NCS in multiple packet transmission with random delays. They modelled multiple packet using Markov's chain process. Recently, Song et al. [64] have proposed the packet loss model using Markov's chain process. The model was validated for a single packet drop as well as successive packet drops. They proposed discrete-time integral sliding mode controller using the proposed model to compensate the effects of packet loss.

1.2.4 Output Feedback Control Algorithms for NCS

The above all literature discusses about design of controllers based on the state information method. The major disadvantage of these controllers is that its performance depends upon the availability of state information [65]. In various applications of NCS such as missile guidance control, aircraft control, chemical industries and automobile sectors, it is not mandatory that all the state information is available. In such cases, it is better to design the controller based on output feedback method. The main advantage of this method is that the performance of controller depends on the availability of output information which is always available. Recently, many researchers have paid much attention to designing the controllers based on output feedback approach in NCS. Mu et al. [66] proposed Luenberger output feedback based controller for discrete-time networked systems. The controller consists of two parts: a state observer that estimates plant states from the output when it is available via network and a model of the plant that is used to generate the control signal when plant output is not available. Similarly, Hespanha and Naghshtabrizi [67] designed observer based Luenberger output feedback to deal with these problems for anticipative and non-anticipative control unit in continuous-time domain. Shi and Yu [56] proved the stability of NCS with random time delays using output feedback method. Zhang and Xia [68] also designed predictive controller that compensates the effect random delays in the presence of matched uncertainty using output feedback approach. Yu and Antsaklis [69] introduced the concept of event triggered for designing output feedback controller in NCS in the presence of time-varying network delays. Zhang et al. [70] designed output feedback sliding mode controller to study the effect of variable time delay in the presence of matched uncertainty. Jungers et al. [71] proved stability of NCSs including global time-varying networked delay. They designed controller based on dynamic output feedback approach dependent on estimation of time-varying delay. Similarly, Wang et al. [72] designed output feedback H_∞ controller for NCSs with packet dropouts, network-induced delays and data drift. They introduced polytopic uncertainty-based data drift to model closed-loop NCSs which include random time delay and packet loss. Qui et al. [73] designed robust output feedback controller for T-S fuzzy-based affine systems with unreliable communication links with multiple packets dropout. Later, Hong et al. [74] designed conventional observer using output feedback approach for wireless NCSs with time-varying network delays and packet dropouts. They modelled wireless NCS using asynchronous dynamic system with assumption that time-varying network delays can be more or less than sampling interval. Yu et al. [75] designed multiple dynamic output feedback controllers for Networked Control Systems in the presence of random time delays and packet loss.

It is worth to mention here that the time required for the data packets to travel from sensor to controller and controller to actuator is defined as total network delay. When such delay is transformed into discrete-time domain, it mostly possesses non-integer type of values. Such network delays in discrete-time domain are defined as fractional delays [76–82]. The networked control system has sensor to controller fractional delay present in the feedback channel and controller to actuator fractional delay present in the forward channel. The nature of both these fractional delays depends on the type of the communication medium. In NCS, when the data packets are exchanged through real-time communication medium, the network delay always have the fractional delay. So, it is important to compensate the effect of deterministic and random type of fractional delay in discrete-time domain at each sampling instant in the presence of packet loss and matched uncertainty.

1.3 Contribution of Book

This book contributes mainly following:

- Firstly, a novel discrete-time sliding surface is proposed using the compensated state information and proposed discrete-time sliding mode control algorithm that encompasses deterministic type network-induced fractional delay and single packet loss. The Thiran approximation technique is used for compensating the fractional delay. There are two types of DSMC proposed, namely switching type and non-switching type. The conditions for stability of the closed-loop system are derived using the Lyapunov approach in both the algorithms. The efficacy of the algorithms is endorsed in simulation as well validated experimentally. Further, the proposed algorithms are compared with conventional sliding mode controller with CAN and Switched Ethernet as network medium.
- The algorithms are further extended for output feedback. A multirate output feedback (MROF) technique is used to estimate the state variables. The proposed

multirate output feedback discrete-time sliding mode controller performance is checked under the networked environment.

- Next, the monograph proposes the discrete-time sliding surface design for the random fractional delay and single packet loss that occur within the sampling period. The random delay is compensated using Thiran's approximation technique in the presence of packet loss situation. The random fractional delay is modelled by Poisson's distribution function, and packet loss is modelled by Bernoulli's function. The closed-loop stability is established using the Lyapunov function. The efficacy of proposed non-switching type of DSMC is endowed by simulation results and also experimentally validated with servo system.
- Further, the proposed algorithms are extended for the random fractional delay with multiple packet loss situation. The multiple packet loss policy is defined for the development of DSMC. The simulation as well as experimental results with various fractional delay situation and matched uncertainties are shown to prove the efficacy of the proposed algorithms.

1.4 Organization of Book

The outline of book is as follows:

- Chapter 1 briefs Networked Control System and the literature survey on various control algorithms proposed in continuous-time as well as discrete-time domain. This chapter also discusses various issue of NCS.
- Chapter 2 discusses the preliminaries of Networked Control System and sliding mode control technique. In this chapter, a basic block diagram of NCS with different types of time delays that affect the performance of the system are discussed. The origin of sliding mode controller in continuous-time domain and discrete-time domain is also briefly discussed. Lastly, the challenges in designing DSMC for NCS are discussed.
- The main contribution of monograph, design of discrete-time sliding mode control for deterministic type fractional delay is discussed in Chap. 3. Firstly, sliding surface is proposed with compensated delay. The network-induced delay is compensated using Thiran's approximation. The discrete-time sliding mode control law is derived using proposed sliding surface with switching type reaching law. Further, the stability of the closed-loop NCS is proved through Lyapunov approach. The efficacy of the proposed algorithm is tested under simulation environment and experimental environment.
- Chapter 4 presents design of non-switching type discrete-time sliding mode controller in the presence of deterministic fractional delay and matched uncertainty. In this chapter, the design of control law is based on sliding surface derived using Thiran's approximation. Further, the stability of the closed-loop NCS is proved through Lyapunov approach that ensures the finite-time convergence of system states within the specified band. The efficacy of the proposed algorithm

is tested under simulation and experimentation with real-time networks like CAN and Switched Ethernet.

- Chapter 5 describes the design of discrete-time sliding mode control using multi-rate output feedback approach with fractional delay compensation. In this chapter, the MROF-based estimator is placed at the plant side using the control input signal and fast measured output. The stability of the closed-loop NCS with derived control law is proved using Lyapunov approach. The simulation results are carried out in the presence of network delay and matched uncertainty in order to prove the effectiveness of proposed algorithm.
- Chapter 6 describes the design of discrete-time sliding mode controller for random communication delay and single packet loss. In this chapter, the compensation of random fractional delay is discussed using Thiran's approximation with packet loss condition. The mathematical models of random fractional delay and single packet loss are derived using stochastic approach. The derived discrete-time sliding mode control law is verified through simulation and implementation results in the presence of random fractional delay, packet loss and matched uncertainty.
- Chapter 7 discusses design of discrete-time sliding mode control for multiple packet transmission. The multiple packet loss policy is defined and used for DSMC design. Further, a second-order disturbance estimator is incorporated at the plant side to estimate the disturbance $O(h^3)$ that occur on the plant. This improves the robustness properties of closed-loop NCS. The efficiency of the proposed algorithm is verified through simulation results.
- The concluding remarks along with future scope and challenges are mentioned in Chap. 8. The final comments and future scope of discrete-time SMC algorithms are discussed in this chapter. Lastly, various challenges are also listed that are still remain unsolved in network control system domain.

References

1. S. Asif, P. Webb, Networked control system - an overview. Int. J. Comput. Appl. **115**(6), 26–30 (2015)
2. R. Gupta, M. Chow, Networked control system: overview and research trends. IEEE Trans. Ind. Electron. **57**(7), 2527–2535 (2010)
3. W. Ling, X. Ding-yu, E. Da-zhi, Some basic issues in networked control systems, in *Second IEEE Conference on Industrial Electronics and Applications* (2007), pp. 2008–2012
4. L. Zhang, H. Gao, O. Kaynak, Network-induced constraints in networked control systems: a survey. IEEE Trans. Ind. Informatics **9**(1), 403–416 (2013)
5. Y. Zhao, X. Sun, J. Zhang, P. Shi, Networked control systems: the communication basics and control methodologies. IEEE Trans. Ind. Informatics **9**(1), 403–416 (2013)
6. R. Luck, A. Ray, Delay compensation in integrated communication and control systems: conceptual development and analysis, in *Proceedings of American Control Conference (ACC90)* (1990), pp. 2045–2050
7. R. Luck, A. Ray, Delay compensation in integrated communication and control systems: implementation and verification, in *Proceedings of American Control Conference (ACC90)* (1990), pp. 2051–2055

8. R. Luck, A. Ray, An observer-based compensator for distributed delays. Automatica **26**(5), 903–908 (1990)
9. R. Luck, A. Ray, Experimental verification of a delay compensation algorithm for integrated communication and control systems. Int. J. Control **59**(6), 1357–1372 (1994)
10. H. Chan, Closed loop control of a system over a communication network with queues. Int. J. Control **62**(3), 493–501 (1995)
11. Z. Yu, H. Chen, Y. Wang, Control of network system with random communication delay and noise disturbance. Control Decis. **15**(5), 518–526 (2000)
12. T.C. Yang, Network control system: a brief survey. IEE Proc. Control Theory Appl. **153**(4), 403–412 (2006)
13. L.A. Montestruque, P.J. Antsaklis, On the model-based control of networked systems. Automatica **39**(10), 1837–1843 (2003)
14. D. Yue, Q. Han, C. Peng, State feedback controller design of networked control systems. IEEE Trans. Circ. Syst.-II **51**(11), 640–644 (2004)
15. D. Kim, J. Ko, P. Park, Stabilization of the asymmetric network control system using a deterministic switching system approach, in *Proceedings of the 41st IEEE Conference on Decision and Control (CDC 02)* (2002), pp. 1638–1642
16. E. Godoy, A. Jose, V. Porto, Using simulation tools in the development of a networked control systems research Platform. ABCM Symp. Ser. Mechatronics **4**(1), 384–393 (2010)
17. Y. Li, S. Yang, Z. Zhang, Q. Wang, Network load minimisation design for dual-rate Internet-based control systems. IET Control Theory Appl **4**(2), 197–205 (2010)
18. S. Vardhan, R. Kumar, Simulations for time delay compensation in networked control systems. J. Sel. Areas in Telecommunications **1**(1), 38–43 (2011)
19. M. Urban, M. M-Blaho, M. Murgas, Foltin, Simulation of networked control systems via truetime. Tech. Computing Prague **1**(1), 1–7 (2007)
20. W. Zhang, M. Branicky, S. Phillips, Stability of networked control systems. IEEE Control Syst. Mag. **1**(1), 84–99 (2001)
21. G. Walsh, H. Ye, L. Bushnell, Stability analysis of networked control systems. IEEE Trans. Control Sys. Technol. **10**(3), 438–446 (2002)
22. Y. Tipsuwan, M.-Y. Chow, Gain scheduler middleware: a methodology to enable existing controllers for networked control and teleoperation-part I: networked control. IEEE Trans. Ind. Electron **51**(6), 1218–1227 (2004)
23. K. Ji, W. Kim, Real-time control of networked control systems via ethernet. Int. J. Control. Autom. Sys. **3**(4), 591–600 (2005)
24. D. Ma, J. Zhao, Stabilization of networked control Systems via switching controllers: an average dwell time approach, in *IEEE Proceedings of World Congress on Intelligent Control* (2006), pp. 4619–4622
25. C. Peng, D. Yue, State feedback controller design of networked control systems with parameter uncertainty and state-delay. Asian J. Control **8**(4), 385–392 (2006)
26. H. Gao, T. Chen, J. Lam, A new delay system approach to network-based control. Automatica **44**(1), 39–52 (2008)
27. G. Liu, Y. Xia, J. Chen, D. Rees, W. Hu, Networked predictive control of systems with random networked delays in both forward and feedback channels. IEEE Trans. Ind. Electron. **54**(3), 1282–1296 (2007)
28. B. del-Muro-Cuellar, M. Velasco-Villa, G. Ferna'ndez-Anaya, O. Jime'nez-Ram'irez, J. A' lvarez-Ram'irez, Observer-based prediction scheme for time-lag processes, in *Proceedings of American Control Conference* (2007), pp. 639–644
29. Y. Sun, J. Xu, Stability analysis and controller design for networked control systems with random time delay, in *The ninth international conference on electronic measurement and instruments*, (2009), pp. 136–141
30. H. Yuhong, S. Yeguo, State feedback controller design of networked control system with time delay in the state derivatives, in *International Conference on Computer Application and System Modeling* (2010), pp. 319–322

31. M. Ono, N. Ban, K. Sasaki, K. Matsumoto, Y. Ishida, Discrete modified smith predictor for an unstable plant with dead time using a plant predictor. Int. J. Comput. Sci. Network Security **10**(9), 80–85 (2010)
32. W. Ridwan, B. Trilaksono, Networked control synthesis using time delay approach: state feedback case. Int. J. Electr. Eng. Informatics **3**(4), 441–452 (2011)
33. M. Vallabhan, S. Srinivasan, S. Ashok, S. Ramaswamy, R. Ayyagari, in *An analytical framework for analysis and design of networked control systems with random delays and packet losses*, IEEE Proceedings of Canadian Conference on Electrical and Computer Engineering (2012), pp. 1–6
34. Y. Hikichi, K. Sasaki, R. Tanaka, H. Shibasaki, K. Kawaguchi, Y. Ishida, A discrete PID control system using predictors and an observer for the influence of a time delay. Int. J. Model. Optim. **3**(1), 1–4 (2013)
35. J. Hu, J. Liang, H. Karimi, J. Cao, Sliding intermittent control for BAM neural networks with delays. Abst. Appl. Anal. **2013**(1), 1–15 (2013)
36. N.T. Cac, N.X. Hung, N.V. Khang, CAN- based networked control systems: a compensation for communication time delays. Am. J. Embedded Sys. Appl. **2**(3), 13–20 (2014)
37. H. Yi, H. Kim, J. Choi, Design of networked control system with discrete-time state predictor over WSN. J. Adv. Computing Networks **2**(2), 106–109 (2014)
38. M.A. Khanesar, O. Kaynak, S. Yin, H. Gao, Adaptive indirect fuzzy sliding mode controller for networked control systems subject to time varying network induced time delay. IEEE Trans.on Fuzzy Sys. **23**(1), 1–10 (2014)
39. R. Saravanakumar, M. Syed Ali, C. Ahn, H. Karimi, P. Shi, Stability of Markovian jump generalized neural networks with interval time-varying delays. IEEE Trans. Neural Network Learning Syst. **99**(1), 1–11 (2016)
40. G. Jacovitti, G. Scarano, Discrete time techniques for time delay estimation. IEEE Trans. Signal Process. **41**(2), 525–533 (1993)
41. J. Nilsson, B. Bernhardsson, B. Wittenmark, Stochastic analysis and control of real time systems with random time delays. Elsevier **2**(3), 13–20 (1997)
42. H. Shousong, Z. Qixin, Stochastic optimal control and analysis of stability of networked control systems with long delay. Automatica **39**(2003), 1877–1884 (2003)
43. P. Zhivoglyadov, R. Middleton, Networked control design for linear systems. Automatica **39**(2003), 743–750 (2003)
44. D. Yue, Q.L. Han, J. Lam, Network-based robust Hinf control of systems with uncertainity. Automatica **41**(1), 999–1007 (2005)
45. Y. Zhao, G. Liu, D. Rees, Integrated predictive control and scheduling co-design for networked control systems. IET Control Theory Appl. **2**(1), 7–15 (2008)
46. X. Gou Controller design based on variable-period sampling approach for networked control systems with random delays, in *IEEE Proceedings of International Conference on Networking, Sensing and Control* (2009), pp. 147–151
47. J. Xiong, J. Lam, Stabilization of networked control systems with a logic ZOH. IEEE Trans. Autom. Control **54**(2), 358–363 (2009)
48. H. Li, H. Yang, F. Sun, Y. Xia, Sliding-mode predictive control of networked control systems under a multiple-packet transmission policy. IEEE Trans. Ind. Electron. **61**(11), 201–221 (2014)
49. P. Guo, J. Zhang, H. Karimi, Y. Liu, M. Lyu, Y. Bo, State estimation for wireless network control system with stochastic uncertainty and time delay based on sliding mode observer. Abstr. Appl. Anal. **2014**(2014), 1–8 (2014)
50. D. Yao, H. Karimi, Y. Sun, Q. Lu, Robust model predictive control of networked control systems under input constraints and packet dropouts. Abstr. Appl. Anal. **2014**(2014), 1–11 (2014)
51. A. Argha, L. Li, W. Su Steven, H. Nguyen, Discrete-time sliding mode control for networked systems with random communication delays, in *Proceedings of American Control Conference* (2015), pp. 6016–6021
52. W. Yang, L. Fan, J. Luo, Design of discrete time sliding mode observer in networked control system, in *IEEE Proceedings of Chinese Control and Decision Conference* (2010), pp. 1884–1887

53. L. Zhang, Y. Shi, T. Chen, B. Huang, A new method for stabilization of networked control systems with random delays. IEEE Trans. Autom. Control **50**(8), 1177–1181 (2005)
54. J. Dong, W. Kim, markov-chain-based output feedback control for stabilization of networked control systems with random time delays and packet losses. Int. J. Control Autom. Sys. **10**(5), 1013–1022 (2012)
55. F. Yang, Z. Wang, Y. Hung, M. Gani, H-infinity control for networked systems with random communication delays. IEEE Trans. Autom. Control **51**(3), 511–518 (2006)
56. Y. Shi, B. Yu, Output feedback stabilization of networked control systems with random delays modeled by Markov Chains. IEEE Trans. Autom. Control **54**(7), 1668–1674 (2009)
57. Y. Ge, Q. Chen, M. Jiang, Y. Huang, Modeling of random delays in networked control systems. J. Control Sci. Eng. **2013**(1), 1–9 (2013)
58. J. Hespanha, P. Naghshtabrizi, Y. Xu, A survey of recent results in networked control systems. Proc. IEEE **95**(1), 138–162 (2007)
59. V. Gupta, B. Hassibi, R. Murray, Optimal LQG control across packet-dropping links. Syst. Control Lett. **56**(2007), 439–446 (2007)
60. J. Wu, T. Chen, Design of networked control systems with packet deopouts. IEEE Trans. Autom. Control **52**(7), 1314–1319 (2007)
61. X. Zhu, G. Yang, State feedback controller design of networked control systems with multiple-packet transmission. Int. J. Control **82**(1), 86–94 (2009)
62. Y. Niu, D. Ho, Design of sliding mode control subject to packet losses. IEEE Trans. Autom. Control **55**(11), 2623–2628 (2010)
63. D. Wen, Y. Gao, H-inifinity controller design for networked control systems in multiple packet transmission with random delays, in *Applied Mechanics and Materials* (2013), pp. 1601–1604
64. H. Song, S. Chen, Y. Yam, *Sliding Mode Control for Discrete-Time Systems With Markovian Packet Dropouts*, IEEE Trans. Cyber. vol. PP, No. 99, pp. 1–11 (2016)
65. A. Mehta, B. Bandyopadhyay, Multirate output feedback based stochastic sliding mode control. J. Dyn. Sys. Meas. Control **138**(12), 124503(1–6) (2016)
66. S. Mu, T. Chu, L. Wang, W. Yu, Output feedback networked control system. Int. J. Autom Computing **1**(1), 26–34 (2004)
67. J. Hespanha, P. Naghshtabrizi Designing an observer-based controller for a network control system, in *IEEE Proceedings of Control and Decision Conference* (2005), pp. 848–853
68. J. Zhang, Y. Xia, Networked predictive output feedback control of networked control systems, in *IEEE Conference on Chinese Control Conference* (2011), pp. 205–210
69. H. Yu, P. J. Antsaklis, Event-Triggered Output Feedback Control for Networked Control Systems using Passivity: Time-varying Network Induced Delaysin, in *IEEE Conference on Decision and Control and European Control Conference* (2011), pp. 205–210
70. J. Zhang, J. Lam, Y. Xia, Output feedback sliding mode control under networked environment. Int. J. Sys. Sci. **44**(4), 750–759 (2013)
71. M. Jungers, E. Castelan, V. Moraes, Ubirajara Moreno, A dynamic output feedback controller for a network controlled system based on delay estimates. Automatica **49**(3), 788–792 (2013)
72. Y. Wang, Q. Han, W. Liu, Modelling and dynamic output feedback controller design for networked control systems, in *IEEE Proceedings of American Control Conference* (2013), pp. 1615–1620
73. J. Qiu, G. Feng, H. Gao, Asynchronous output-feedback control of networked nonlinear systems with multiple packet dropouts: TÜS fuzzy affine model-based approach. IEEE Trans. Fuzzy Sys. **19**(6), 1030–1040 (2013)
74. Z. Hong, J. Gao, N. Wang, Output-feedback controller design of a wireless networked control system with packet loss and time delay. Mathe. Problems Eng. **2014**, 1–7 (2014)
75. J. Yu, M. Yu, H. Shi, B. Liu, Y. Gao, Design of multiple dynamic output feedback controllers for networked control systems, in *IEEE Proceedings of Chinese Control Conference* (2014), pp. 5818–5823
76. D. Shah, A. Mehta, Discrete-time sliding mode control using Thiran's delay approximation for networked control system, in *43rd Annual Conference on Industrial Electronics (IECON-17)* (Nov. 2017)

77. D. Shah, A.J. Mehta, Design of robust controller for networked control system, in *Proceedings of IEEE International Conference on Computer, Communication and Control Technology* (Sept. 2014), pp. 385–390
78. D. Shah, A. Mehta, Output feedback discrete-time networked sliding mode control, in *IEEE Proceedings of Recent Advances in Sliding Modes (RASM)* (2015), pp. 1–7
79. A.J. Mehta, B. Bandyopadhyay, Frequency-shaped and observer- based discrete- time sliding mode control, in *SpringerBriefs in Applied Sciences and Technology* (2015)
80. D. Shah, A. Mehta, Discrete-time sliding mode controller subject to real-time fractional delays and packet losses for networked control system. Int. J. Control Autom. Syst. (IJCAS) **15**(6), 2690–2703 (2017)
81. D. Shah, A. Mehta, Fractional delay compensated discrete-time SMC for networked control system, in Digital Communication Networks (DCN). Elsevier **2**(3), 385–390 (2016)
82. D. Shah, A. Mehta, Multirate output feedback based discrete-time sliding mode control for fractional delay compensation in NCSs, in *IEEE Conference on Industrial Technology (ICIT-2017)* (2017) pp. 848–853

Chapter 2
Preliminaries of Sliding Mode Control and Networked Control System

Abstract In this chapter, we introduce the concept of Networked Control System (NCS), the irregularities such as time delay and packet loss that occurs in the NCS. We also discussed the various control methods available for NCS. This chapter also presents the concept of sliding mode control along with the literature survey on SMC for NCS.

Keywords Network control system · Time delay · Packet loss
Discrete-time sliding mode control

2.1 Brief Review of Sliding Mode Control (SMC) Technique

2.1.1 Origin of Sliding Mode Control

The concept of variable structure control (VSC) was introduced by Emelyanov group in late 1950s [1]. The main idea of VSC was to switch between the various control structures according to the evaluation of the system states. The concept of VSC can be understood through the following example.
Consider two constituent systems given as:

$$\ddot{x} = -a_1 x, \tag{2.1}$$

$$\ddot{x} = -a_2 x, \tag{2.2}$$

where $0 \prec a_2 \prec a_1$. The phase portraits of the systems are shown in Figs. 2.1 and 2.2, respectively. It can be observed that both the systems are oscillatory in nature and are unstable. But when both structures are switched at appropriate time, then combined system is asymptotically stable. The combined response of both the system is shown in Fig. 2.3. Thus it can be noticed that the property not present in any of the system is obtained by VSC.

© Springer Nature Singapore Pte Ltd. 2018

D. H. Shah and A. Mehta, *Discrete-Time Sliding Mode Control for Networked Control System*, Studies in Systems, Decision and Control 132, https://doi.org/10.1007/978-981-10-7536-0_2

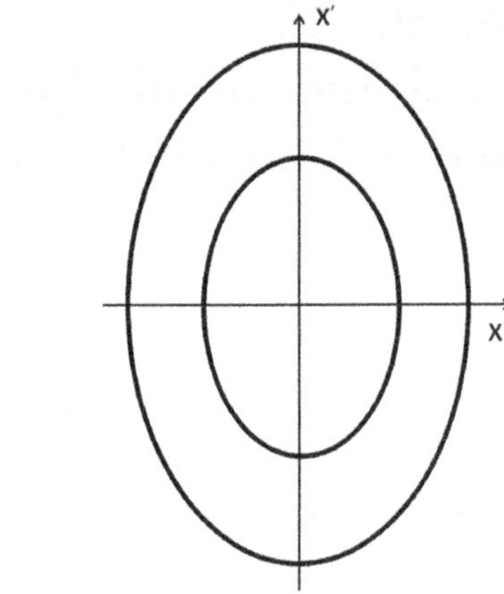

Fig. 2.1 State trajectories of system in mode-I

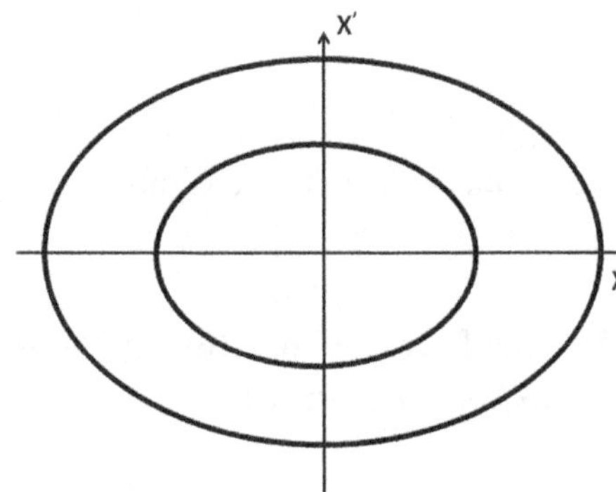

Fig. 2.2 State trajectories of system in mode-II

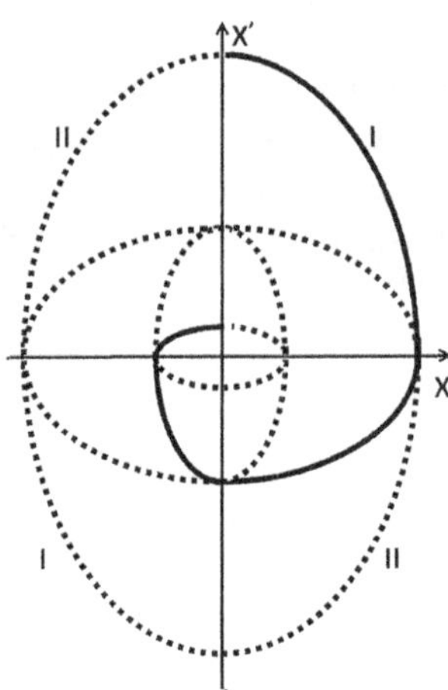

Fig. 2.3 Combined system response

While carrying out research on VSC, some of the researchers [2–5] made an unusual observation that switching between two or more unstable control structures may result in stable control system. They introduced the notions of variable structure control, variable structure system, and a new control idea called sliding mode control came into existence. The sliding mode control, or SMC, is a nonlinear control method that alters the dynamics of a nonlinear system by application of a discontinuous control signal that forces the system to "slide" along a cross-section of the system's normal behaviour. Hence, sliding mode control is a variable structure control method. The multiple control structures are designed so that trajectories always move towards an adjacent region with a different control structure, and so the ultimate trajectory will not exist entirely within one control structure. Instead, it will slide along the boundaries of the control structures. The motion of the system as it slides along these boundaries is called a sliding mode, and the geometrical locus consisting of the boundaries is called the sliding surface. The conventional example of sliding mode is second-order relay system which is given by,

$$\ddot{x} + a_2\dot{x} + a_1 x = u, \tag{2.3}$$

$$u = -M_s sign(s), \tag{2.4}$$

$$s = cx + \dot{x}, \tag{2.5}$$

where a_1, a_2, M_s and c are constants. From Eq. (2.4), it can be noticed that the control input in the second order system might take only two values M_s and $-M_s$ and causes discontinuities on the straight line $s = 0$ in the state plane $(x, \dot{x})$. Figure 2.4 shows the response of the sliding mode for the specified example with $a_1 = a_2 = 0$. From the result it can be observed that in the neighbourhood segment mn on the switching line $s = 0$, the trajectories of the system (2.3) run in the opposite directions which lead to the appearance in sliding mode along this line [6]. Thus, Eq. (2.5) can be interpreted as sliding mode equation. Further, it can be noticed that the order of sliding mode equation is less than the original system (2.3). Thus, it can be said that the sliding mode does not depend on plant dynamics but it is determined by parameter c only.

Fig. 2.4 Sliding mode for relay system

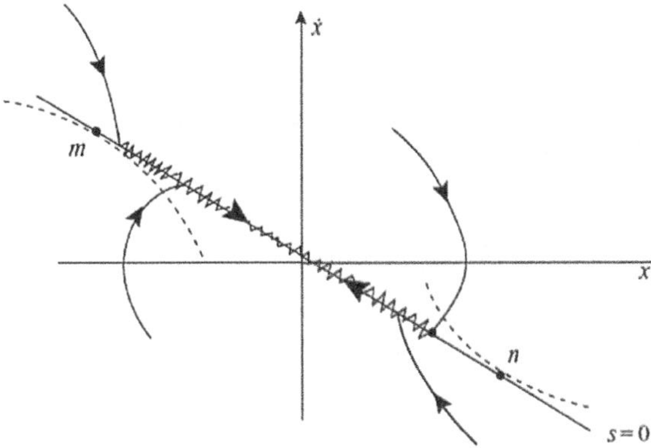

It is worth to point out that for desired performance of the closed-loop system not only the sliding mode controllers have to be properly designed but the switching rule should also be chosen properly. Based on the appropriate selection of switching rule and designed controller, the system states will drive onto the predefined sliding surface in finite time. This ensures stability of sliding motion on the surface, and desired dynamic characteristics of the systems are achieved.

Due to its lower sensitivity towards plant parameters variations and external disturbances, it eliminates the necessity of exact modelling. It allows the decoupling of overall system motion into partial components of lower dimensions. This reduces the complexity of feedback design. In sliding mode control, the control actions are function of discontinuous state which can be easily implemented using conventional power converters. Due to such properties, SMC has been proved applicable to a wide range of applications such as robotics, electrical drives, electrical generators, motion control, process control and Networked Control System.

2.1.2 Continuous-Time Sliding Mode Control

In sliding mode control, the control law consists of two important phases: (i) reaching phase and (ii) sliding mode phase. When the system state is driven from any initial state to reach the switching manifold in finite time, then such phase is called reaching phase, and when the system is induced into the sliding motion on the switching manifolds, then such phase is defined as sliding mode phase. Figure 2.5 shows the phases of sliding mode control with s as continuous-time sliding function given by:

$$s = \{x \in X | s(x, t) = 0\}. \tag{2.6}$$

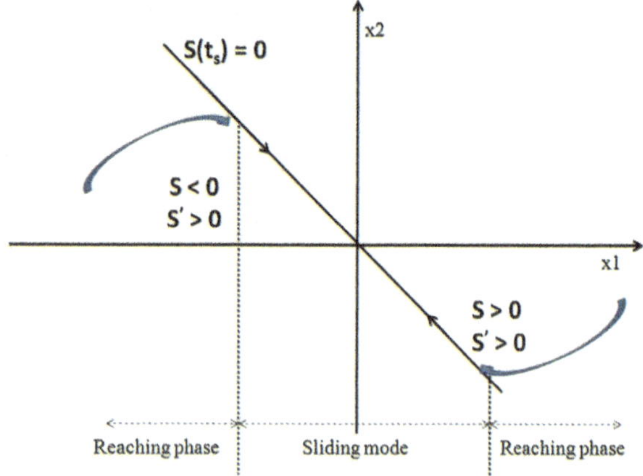

Fig. 2.5 Phases of sliding mode control

In order to induce the sliding mode, the following properties should exist: (i) the system stability should be restricted to the sliding surface, and (ii) the sliding mode should start within a finite time. The sufficient condition for the sliding motion to slide on the given surface is given by:

$$s\dot{s} \prec 0, \qquad (2.7)$$

where s is the sliding surface and $\dot{s}$ is rate of change of distance from the sliding surface. The condition specified in Eq. (2.7) is called a reachability condition. The reachability condition is not sufficient for the sliding mode. The main drawback of condition mentioned in (2.7) is that $s(t)$ takes infinite time to reach on the sliding surface. Thus, to overcome this drawback another condition is defined as:

$$s\dot{s} \prec -\eta|s|, \eta \succ 0. \qquad (2.8)$$

This condition is known as "η-reachability" condition that ensures the finite-time convergence to $s = 0$.

As discussed earlier, the designing of the sliding mode controller includes reaching law design, sliding surface design and control law design. Let us consider the continuous-time system given by:

$$\dot{x}(t) = Ax(t) + Bu(t), \qquad (2.9)$$

$$y(t) = Cx(t). \qquad (2.10)$$

The dynamics of the sliding function can be expressed in the form of constant rate reaching law as:

$$\dot{s} = -k_s sgn(s), k_s \succ 0. \qquad (2.11)$$

Let the sliding surface be given by:

$$s(t) = C_s x(t), \qquad (2.12)$$

where C_s is the sliding gain that can be computed using pole-placement method based on the proper selection of poles or LQR method based on the proper selection of Q and R matrices.

Thus, the control law for the system (2.9, 2.10) is derived by using the condition $\dot{s} = 0$ as:

$$u(t) = -(C_s B)^{-1}[AC_s x(t) + k_s sgn(s)]. \qquad (2.13)$$

The reaching laws proposed in the literatures [7, 8] are,

- Constant proportional rate

$$\dot{s} = -qs - k_s sgn(s), q \succ 0 \qquad (2.14)$$

- Power rate reaching law

$$\dot{s} = -k_s |s|^\iota sgn(s), 0 \prec \iota \prec 1 \tag{2.15}$$

- Exponential reaching law

$$\dot{s} = -\frac{k_s}{N(s)} sgn(s), \tag{2.16}$$

where $N(s) = \delta_0 + (1 - \delta_0)e^{-\iota |s|^{p_0}}$, δ_0 is strictly positive offset less than 1 and p_0 is a strictly positive integer.

The main limitation of continuous-time SMC is that once the closed-loop system states reach on the sliding surface, a discontinuous control term switches with high frequency which results in chattering phenomenon. The chattering is caused due to various reasons such as switching time delay, controller computation delay, dynamics of plant elements such as actuator and sensor. In practical applications, this phenomenon is not desirable as it affects the performance of plant, while in electrical and mechanical applications, it causes high heat losses and generates wear and tear of the moving parts of machines. In discrete-time sliding mode control, generally low switching frequencies are required because of limited sampling frequency due to which it becomes more useful for practical implementation. Moreover, in discrete-time SMC the computation of control signal is done at each sampling interval and remains constant for that period. So due to this the system state trajectory is unable to move along the sliding surface but follows the zigzag motion about the sliding surface defined as quasi-sliding mode motion.

2.1.3 Discrete-Time Sliding Mode Control

The concept of discrete-time sliding mode control is first introduced by [9]. In his work, he proved that the sliding motion in discrete-time is more accurate than continuous-time sliding mode control. Later on, [2, 10–15] and many others extended their work in discrete-time sliding mode control.

The designed procedure of DSMC includes: (i) design of sliding surface, (ii) reaching law and (iii) control law that steers the system states to slide along predefined sliding surface over a finite interval of time. In discrete-time SMC, Gao's [12] introduced the concept of switching function in the reaching law that causes the system states to move towards the vicinity of the origin but cannot get arbitrarily closed to the origin, while Bartolini and Bartoszewicz [11, 13] designed the reaching law without considering the switching function. They considered that discrete-time control is naturally discontinuous in nature and thus may not require an explicit discontinuity in the control law. The reaching law proposed by them causes the system states to get arbitrarily closed to the origin.

Various state-based discrete-time control algorithms are designed using different reaching laws available in [9–14], respectively.

In order to derive the discrete-time control algorithms, let us consider the continuous-time SISO system as:

$$\dot{x}(t) = Ax(t) + Bu(t), \tag{2.17}$$

$$y(t) = Cx(t), \tag{2.18}$$

where $x \in R^n$ represents system state vector, $u \in R^m$ represents control input, $y \in R^p$ represents system output and $A \in R^{n \times n}$, $B \in R^{n \times m}$ and $C \in R^{p \times n}$ are the matrices of appropriate dimensions.

Let the system (2.17) and (2.18) be discretized at h sampling interval given by,

$$x(k+1) = Fx(k) + Gu(k), \tag{2.19}$$

$$y(k) = C(k). \tag{2.20}$$

As discussed earlier, the design of sliding mode control algorithms involves the design of sliding surface and reaching law. Let the discrete-time sliding surface be given by:

$$s(k) = C_s x(k). \tag{2.21}$$

Various researchers have proposed famous reaching laws in discrete-time domain as listed below.

- Sarpturk's Reaching Law: The reaching law proposed by [10] is the direct discretization of continuous-time sliding mode given by:

$$|s(k+1)| \prec |s(k)|. \tag{2.22}$$

Here, the sliding surface is always directed towards the surface and also the norm of $|s(k)|$ monotonically decreases. The reaching law can be written in other way as,

$$(s(k+1) - s(k))sgn(s(k)) \prec 0, \tag{2.23}$$

$$(s(k+1) - s(k))sgn(s(k)) \succ 0. \tag{2.24}$$

The first condition (2.23) indicates that the closed-loop system state trajectories should move towards the direction of sliding surface, and the second condition (2.24) indicates that the closed-loop system state trajectories are not allowed to go too far in that direction. Thus observing conditions (2.23) and (2.24) will lead to lower and upper bounds for control actions. The control law proposed in [10] is given as,

$$u(k) = -k_t(x, s)x(k), \tag{2.25}$$

where k_t is the gain given by:

$$k_t(x, s) = \begin{cases} k_t^+; & \text{when } x(k)s(k) > 0 \\ k_t^-; & \text{when } x(k)s(k) < 0 \end{cases}$$

where k_t^+ and k_t^- represent the coefficients of each upper bound and lower bound of control action that can be determined by evaluating the condition (2.23) and (2.24), respectively.

- Gao's Reaching Law: The switching-based reaching law proposed by [12] is given as,

$$s(k + 1) = (1 - qh)s(k) - \varepsilon h sgn(s(k)), \tag{2.26}$$

where h is the sampling interval satisfying $h > 0, q, \varepsilon > 0$ and $1 - qh > 0$. Using reaching law (2.26), the switching-based control law for system (2.19, 2.20) is computed as,

$$u(k) = -(C_s G)^{-1}[FC_s x(k) - (1 - qh)s(k) + \varepsilon h sgn(s(k))]. \tag{2.27}$$

From above Eq. (2.27), it can be noticed that there are two parameters q and ε in control law for tuning the response. The discrete-time sliding mode control law proposed in (2.27) should achieve the following performances [12]. (i) Starting from any initial state, the trajectory will move monotonically towards the switching plane and cross it in finite time. (ii) Once the trajectory has crossed the switching plane, it will cross the plane again in every successive sampling period, resulting in a zigzag motion about the switching plane, and (iii) The size of each successive zigzag step is non-increasing, and the trajectory stays within a specified band. The reaching law in Eq. (2.26) states that the state vector always moves towards the quasi-sliding mode band given as:

$$s(k) \leq \frac{\varepsilon h}{1 - qh}. \tag{2.28}$$

From Eqn. (2.28), it can be observed that ε is directly proportional to quasi-sliding mode band. Thus if the value of ε is too large, then the system will have high overshoots and could also increase the transient response. While on the other hand the value of qh should be less than unity, otherwise it will speed up the transient response.

- Bartoszewicz's Reaching Law: [13, 14] proposed non-switching reaching law as,

$$s(k + 1) = d(k) - d_0 + s_d(k + 1), \tag{2.29}$$

where $d(k)$ is the unknown disturbance, d_0 is the mean value of disturbance $d(k)$ and μ_d is minimum deviated disturbance with d_u as upper bound and d_l as lower bound. Also, d_0 and d_2 are given by

$$d_0 = \frac{d_l + d_u}{2} \text{ and } d_2 = \frac{d_u - d_l}{2}$$

$s_d(k)$ is an priori known function such that the following applies:
- If $s(0) \succ 2d_2$ then,

$$s_d(0) = s(0)$$
$$s_d(k)s_d \geq 0 \text{ for any } k \geq 0$$
$$s_d(k) \geq 0 \text{ for any } k \geq k'$$
$$|s_d(k + 1)| \prec |s_d(k)| - 2d_2 \text{ for any } k \leq k'$$

These relations state that $s_d(k)$ converges monotonically from its initial position to the origin of the state space in a finite time. Moreover, in each control step the hyperplane moves by the distance greater than $2d_2$. This, together with Eq. (2.29), states that even in the case of worst combination of disturbance the reaching condition is satisfied.
- If $s(0) \prec 2d_2$ then $s_d(k) = 0$ for any $k \geq 0$.

The k' in the above relations is a positive integer user-defined constant which provides the faster convergence rate of the system and magnitude of the control signal $u(k)$. The control law computed using the reaching law (2.29) for the system (2.19, 2.20) is given as,

$$u(k) = (C_s G)^{-1}[C_s F x(k) + d_0 - s_d(k + 1)]. \tag{2.30}$$

The reaching law in Eq. (2.29) states that the system state vector always move towards the QSMB for any $k \geq k'$ such that:

$$|s(k)| = |d(k - 1) - d_0| \leq d_2. \tag{2.31}$$

- Bartoszewicz's Reaching Law: [16] proposed the other reaching law that provides the faster convergence of sliding variable without increasing the amplitude of the control signal. The reaching law is given as,

$$s[(k + 1)h] = \{1 - q[s(kh)]\}s(kh) - \hat{S}(kh) - d(kh) + d_1 + S_1, \tag{2.32}$$

where $\hat{S}(kh)$ represents the model uncertainty on sliding variable evolution and $d(kh)$ represents the effect of disturbance on this variable. Further, S_1 and d_1 represent the mean values of $\hat{S}(kh)$ and $d(kh)$, respectively, given as,

$$S_1 = \frac{S_u + S_l}{2}, \tag{2.33}$$

$$d_1 = \frac{d_u + d_l}{2}, \tag{2.34}$$

where S_u and S_l are upper and lower bounds of S_1 and d_u and d_l are upper and lower bounds of d_1.

The convergence rate factor of $q[s(kh)]$ in Eq. (2.32) is given as,

$$q[s(kh)] = \frac{\psi}{\psi + |s(kh)|}, \tag{2.35}$$

where ψ is designer's constant satisfying $\psi > S_2 + d_2$, where S_2 and d_2 are the greatest possible deviation of $\hat{S}$ and d. They are represented as,

$$S_2 = \frac{S_u - S_l}{2}, \tag{2.36}$$

$$d_2 = \frac{d_u - d_l}{2}. \tag{2.37}$$

- The control law computed using the reaching law (2.32) for the system (2.19, 2.20) is given as,

$$u(k) = (C_s G)^{-1}[C_s F x(k) + \{1 - q[s(kh)]\}s(kh) + S_1 + d_1]. \tag{2.38}$$

The reaching law in Eq. (2.32) states that the system states always move towards the QSMB for any $k > k_0$ such that:

$$|s(kh)| \leq \frac{\psi(S_2 + d_2)}{\psi - (S_2 + d_2)}. \tag{2.39}$$

- Exponential Reaching Law: [17] proposed exponential reaching law that provides faster convergence with smaller width of quasi-sliding mode domain (QSMD). The ultimate magnitude of the QSMD in proposed method is of the order $O(T^3)$. The reaching law is given as:

$$s(k+1) = (1 - qh)\phi(k)s(k) - \frac{\lambda_s}{\phi(k)}sgn(s(k)) + \zeta(k), \tag{2.40}$$

where λ_s is the switching gain,
$\zeta(k) = C_s[d(k) - 2d(k-1) + d(k-2)]$,
$\phi(k) = \delta + (1 - \delta)e^{\varphi|s(k)|^\gamma}$,
$\gamma > 0, 0 \prec \delta \prec 1, \lambda > 0$ and $0 \prec 1 - qh \prec 1$.
- The control law computed using the reaching law (2.40) for the system (2.19, 2.20) is given as,

$$u(k) = (C_s G)^{-1}[C_s F x(k) - (1 - qh)\phi(k)s(k) + 2Cd(k-1) - Cd(k-2)$$
$$+ \frac{\lambda_s}{\phi(k)}sgn(s(k))] \tag{2.41}$$

The reaching law in Eq. (2.40) states that the system states always move towards the QSMB for any $k > k_0$ such that:

$$|s(k)| \prec \lambda_s + \zeta. \qquad (2.42)$$

2.1.4 Advantages of DSMC Over CSMC

- With the increase in use of digital computers and microcontrollers for the implementation of control algorithms, a discrete-time model of the system is justified.
- To validate the better performance of the system, it is better to implement the discrete-time algorithms rather continuous-time algorithms.
- In continuous-time sliding mode control due to high-frequency switching chattering takes place which may cause damage to the system. So, it is not used for all practical applications. While in discrete-time sliding mode control, relatively low switching frequencies are required so DSMC algorithm is more practical to implement.
- When continuous-time algorithms are implemented using digital controllers for implementation, the chattering generated around the sliding mode and stability of the sliding mode is compromised.
- A large class of discrete-time systems is computer controlled, and information about the system measurements is available only at specific time instances, and control inputs can only be changed at these time instances, e.g. biological systems, thyristor, radar system, economic systems.

2.2 Brief Overview on Networked Control System (NCS)

NCS is mainly classified into three structures: (i) shared network structure, (ii) hierarchical network structure and (iii) direct network structure [18, 19]. Figure 2.6 shows the block diagram of typical network control system with direct network structure. In this structure, the sensor, plant and actuator are connected to controller through communication network. The communication network in NCS transfers the data in the form of packets. The thick lines indicate the continuous-time data signal, while the dotted lines indicate discrete-time data signals. As shown in Fig. 2.6, the data signal transmitted from sensor to controller through the network is called as feedback channel while the control signal transmitted from controller to actuator through the same network is called as forward channel. In NCS, most of the applications are based on time-sensitive parameter. So, in such cases, if the network delay increases beyond its tolerance limit the plant or the device can either be damaged or gives inferior performance [20–22].

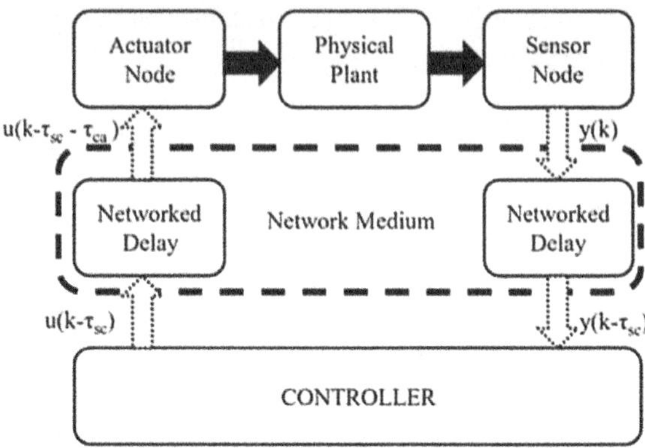

Fig. 2.6 Block diagram of network control system

2.2.1 Network Irregularities in NCS

In NCS, there are four types of time delay: (i) sensor to controller delay, (ii) controller to actuator delay, (iii) computational delay and (iv) processing delay. The sensor to controller delay is present in the forward channel, and controller to actuator delay is present in the feedback channel. The time required for the data to travel from sensor to controller is called a sensor to controller network delay. And similarly the time required for control signal to travel from controller to actuator is called a controller to actuator delay. The combination of both these delays is defined as network delays. The computational delay and processing delay are caused due to the presence of sensor, controller and actuator. So they are also defined as the system delays [23].

The combination of system delays and network delays is defined as total delay, and mathematically it is represented as:

$$\tau_t = \tau + \tau_p, \tag{2.43}$$

where τ is the total network delay and τ_p is the system delays.

The mathematical representation of total network delay and system delay is given as:

$$\tau = \tau_{sc} + \tau_{ca}, \tag{2.44}$$

$$\tau_p = \tau_{sp} + \tau_{cp} + \tau_{ap}, \tag{2.45}$$

where τ_{sc} is sensor to controller delay, τ_{ca} is controller to actuator delay, τ_{ap} is actuator processing delay, τ_{cp} is controller computational delay and τ_{sp} is sensor processing delay.

In NCS, it is always assumed that the effect of system delays (τ_p) is negligible compared to network delays (τ). So, Eq. (2.43) can be written as,

$$\tau_t = \tau. \tag{2.46}$$

Thus, from above Eq. (2.46), it can be noticed that when real-time networks are not considered in NCS the total network delays are always equal to total delay. These network delays can either be deterministic or random in nature.

The nature of the network delays (τ) depends upon the configuration of networks, while the system delays (τ_p) are always deterministic in nature. When the communication takes place using the concept of lease line, then network delays (τ) are deterministic in nature. And when the communication medium is shared by large number of devices, then network delays (τ) are random in nature [23].

A natural assumption on network delays (τ) can be made as,

$$\tau \prec h, \tag{2.47}$$

Or

$$\tau_l \leq \tau \leq \tau_u. \tag{2.48}$$

where h is sampling interval, τ_l is lower bound of delay and τ_u is upper bound of delay.

Observing condition (2.47) and (2.48), it can be concluded that the network delay should always be bounded. During transmission if the packet gets delayed or fails to arrive at the destination within the specified condition (2.47) and (2.48), then such packets are considered as lost packets within the network.

In NCS, the packet loss occurs either in the forward channel or feedback channel. The packet loss is broadly classified into two different categories based on the distance of communication: (i) If the distance of communication is shorter, the data transfer in NCS takes place in the form of frames. Such frame when lost during transmission is treated as single packet loss. (ii) If the distance of communication is longer, the same frame is breakdown in the form of small packets. Such packets when lost during transmission are defined as multiple packet loss.

2.3 Challenges in Designing DSMC for NCS

There are various challenges in network control system with sliding mode control that has not been explored. Some of them are listed below:

- In the literature, the researchers have proposed various SMC algorithms with time delay compensation. But in all the papers, it is assumed that the delays are multiples of sampling interval. However, when the communication is carried through real-time network the delays always have fractional behaviour. Thus, there is need to design SMC for fractional delay in discrete-time domain.
- Furthermore, it can also be noticed that none of the literatures on SMC discusses the designing of sliding surface that compensates the effect of fractional delay. So

there is a need of designing the sliding surface such that it compensates the effect
of fractional delay.

- The compensation of network delays and packet loss is done at the controller side.
  None of the researcher has tried to compensate their effects in the sliding surface.
- Till now the sliding mode controllers are designed based on the presence of multiple
  packet transmission, but the compensation of multiple packets loss with sliding
  mode control in discrete-time domain is still an open research problem in NCSs.
- Various researchers have tried to explore their work on designing the sliding mode
  controllers based on conventional output feedback method in the presence of network delay and packet loss. But the compensation algorithm based on output
  feedback method in the field of NCSs has not been much explored.
- The designing of the higher order sliding modes in the presence of network non-
  idealities (such as random time delay, packet loss and matched uncertainty) is still
  an open research area in NCSs.

2.4 Sliding Mode Control with NCS

Many researchers have proposed different controller design methodologies in discrete-
time as well continuous-time domain for NCSs such as gain scheduler middleware
[24], state feedback [25, 26], adaptive controller [27], Smith predictive [28], back-
stepping controller [29], fuzzy-based sliding mode control [30], H_∞ [31], sliding
mode controller (SMC) [32, 33]. Among all these controllers, SMC has been an
active research area in the field of NCS because it has ability to reject the disturbances which makes it more robust.

In past few decades, many researchers have tried to implement the SMC algorithm in various ways to compensate the effects of network delay and packet loss in
NCSs. Wang et al. [34] proposed variable structure control algorithm for time delay
system. They proposed the controller based on the concept of Smith predictor. Gao
[35] designed integral type of sliding mode control in continuous-time domain in
the presence of variable time delay and matched uncertainty. They proposed sliding
mode compensator in reaching law that compensates the effect of variable network
delay. Goyal et al. [36] designed fuzzy-based sliding mode control in continuous-
time domain. They considered the state-based delay rather than control input delay.
Recently, Khanesar et al. [30] proposed indirect fuzzy-based sliding mode control
in continuous-time domain for NCS. The effect of random time delay was compensated using Pade approximation, and packet loss was compensated using probability
distribution function.

With the rapid development in digital controllers, various researchers have con-
tributed their work in discrete-time domain. The main advantage of designing the
controllers in discrete-time domain is that the effects of control signal can be observed
very clearly at each sampling instant. Moreover, in case of NCS since the communication is carried out in digital signal form so it is better to design SMC in discrete-time
rather than continuous-time domain. Niu and Ho [37] designed sliding mode control

that compensates the effect of single packet loss using probability distribution function. Xiaojuan et al. [38] designed fuzzy-based sliding mode control for NCS and studied the effect of time delay using Ethernet as a network medium in discrete-time domain. Yin et al. [39] designed adaptive based sliding mode control in the presence of variable time delay and uncertainties. They considered sensor to controller delay and transformed the control signal into non-delayed form using fuzzy fusion system in discrete-time domain. Zhang et al. [40] designed output feedback-based sliding mode controller to study the effect of variable time delay in the presence of matched uncertainty. Li et al. [41] designed sliding mode controller using Kalman predictor in the presence of multiple packet transmission. The Kalman predictor was used to estimate the integer type of network delays. Argha et al. [42] proposed discrete-time sliding mode control that compensates the effect of random time delay and packet loss using Bernoulli's distribution. They considered integer type of network delay in discrete-time domain. Recently, Song [43] designed integral sliding mode controller that compensates the effect of single as well as multiple packet loss using Markov's chain process.

2.5 Conclusion

In this chapter, we introduced the basic concept of NCS and its configurations. The various irregularities of NCS are also discussed. The sliding mode control (SMC) technique for NCS is also explained, and at last challenges for designing the SMC for NCS are also discussed.

References

1. S. Emelyanov, S. Korovin, Applying the principle of control by deviation to extend the set of possible feedback types. Sov. Phys. Dokl. **26**(6), 562–564 (1981)
2. V.I. Utkin, Sliding mode control design principles and applications to electric drives. IEEE Trans. Ind. Electron. **40**(1), 23–36 (1993)
3. C. Edwards, S.K. Spurgeon, Robust output tracking using a sliding mode controller/observer scheme. Int. J. Control **64**(5), 967–983 (1996)
4. C. Drakunov, D. Izosimov, A. Loukianov, V.A. Utkin, V.I. Utkin, Block control principle, I and II. Autom. Remote Control **51**(5), 601–609 (1990)
5. K. Furuta, Sliding model control of a discrete system. Syst. Control Lett. **14**, 144–150 (1990)
6. A. Sabanovic, L. Fridman, S. Spurgeon, in *Variable Structure Systems from Principles to Implementation* (IET control engineering, 2004)
7. C. Fallaha, M. Saad, H. Kanaan, K. Haddad, Sliding mode robot control with exponential reaching law. IEEE Trans. Ind. Electron. **58**(2), 600–610 (2016)
8. A.J. Mehta, B. Bandyopadhyay, in *"Frequency-Shaped and Observer-Based Discrete-Time Sliding Mode Control"*. SpringerBriefs in Applied Sciences and Technology, 2015
9. C. Milosavljevic, General conditions for the existence of a quasisliding mode on the switching hyperplane in discrete variable structure systems. Autom. Remote Control **46**, 307–314 (1985)

10. S. Sarpturk, Y. Istefanopulos, O. Kaynak, On the stability of discrete-time sliding mode control systems. IEEE Trans. Autom. Control **32**(10), 930–932 (1987)
11. G. Bartolini, A. Ferrara, V. Utkin, Adaptive sliding mode control in discrete-time systems. Automatica **32**(5), 773–796 (1995)
12. W. Gao, Y. Wang, A. Homaifa, Discrete-time variable structure control systems. IEEE Trans. Ind. Electron. **42**(2), 117–122 (1995)
13. A. Bartoszewicz, Remarks on discrete-time variable structure control systems. IEEE Trans. Ind. Electron. **43**, 235–238 (1996)
14. A. Bartoszewicz, Discrete-time quasi-sliding-mode control strategies. IEEE Trans. Ind. Electron. **45**, 633–637 (1998)
15. W. Su, S.V. Drakunov, U. Ozguner, An $O(T^2)$ boundary layer in sliding mode for sampled-data systems. IEEE Trans. Autom. Control **45**(3), 482–485 (2000)
16. A. Bartoszewicz, P. Lesniewski, Reaching law approach to the sliding mode control of periodic review inventory systems. IEEE Trans. Autom. Sci. Eng. **11**(3), 810–817 (2014)
17. H. Ma, J. Wu, Z. Xiong, A novel exponential reaching law of discrete-time sliding-mode control. IEEE Trans. Ind. Electron. **64**(5), 3840–3850 (2017)
18. D. Shah, A. Mehta, Discrete-time sliding mode control using Thiran's delay approximation for networked control system, in *43rd Annual Conference on Industrial Electronics (IECON-17)*, pp. 3025–3031, Nov 2017. ISBN 978-1-5386-1126-5
19. D. Shah, A.J. Mehta, Design of robust controller for networked control system, in *Proceedings of IEEE International Conference on Computer, Communication and Control Technology*, pp. 385–390, Sept 2014
20. D. Shah, A. Mehta, Output feedback discrete-time networked sliding mode control. in *IEEE Proceedings of Recent Advances in Sliding Modes (RASM)*, pp. 1–7, 2015
21. D. Shah, A. Mehta, Discrete-time sliding mode controller subject to real-time fractional delays and packet losses for networked control system. Int. J. Control Autom. Syst. (IJCAS) **15**(6), 2690–2703 (December 2017)
22. D. Shah, A. Mehta, Fractional delay compensated discrete-time SMC for networked control system. Digit. Commun. Networks (DCN) **2**(3), 385–390 (Dec 2016), Elsevier
23. D. Shah, A. Mehta, Multirate output feedback based discrete-time sliding mode control for fractional delay compensation in NCSs, in *IEEE Conference on Industrial Technology (ICIT-2017)*, pp. 848–853, 2017
24. Y. Tipsuwan, M.-Y. Chow, Gain scheduler middleware: a methodology to enable existing controllers for networked control and teleoperation-Part I: networked control. IEEE Trans. Ind. Electron. **51**(6), 218–1227 (2004)
25. C. Peng, D. Yue, State feedback controller design of networked control systems with parameter uncertainty and state-delay. Asian J. Control **8**(4), 385–392 (2006)
26. H. Gao, T. Chen, J. Lam, A new delay system approach to network-based control. Automatica **44**(1), 39–52 (2008)
27. M. Vallabhan, S. Srinivasan, S. Ashok, S. Ramaswamy, R. Ayyagari, An nalytical framework for analysis and design of networked control systems with random delays and packet losses, in *IEEE Proceedings of Canadian Conference on Electrical and Computer Engineering*, 2012, pp. 1–6
28. Y. Hikichi, K. Sasaki, R. Tanaka, H. Shibasaki, K. Kawaguchi, Y. Ishida, A discrete PID control system using predictors and an observer for the influence of a time delay. Int. J. Model. Optim. **3**(1), 1–4 (2013)
29. H. Yi, H. Kim, J. Choi, Design of networked control system with discrete-time state predictor over WSN. J. Adv. Comput. Netw. **2**(2), 106–109 (2014)
30. M.A. Khanesar, O. Kaynak, S. Yin, H. Gao, Adaptive indirect Fuzzy sliding mode controller for networked control systems subject to time varying network induced time delay. IEEE Trans. Fuzzy Syst. **23**(1), 1–10 (2014)
31. D. Yue, Q.L. Han, J. Lam, Network-based robust Hinf control of systems with uncertainity. Automatica **41**(1), 999–1007 (2005)

32. A. Mehta, B. Bandyopadhyay, Frequency-shaped sliding mode control using output sampled measurements. IEEE Trans. Ind. Electron. **56**(1), 28–35 (2009)
33. A. Mehta, B. Bandyopadhyay, Multirate output feedback based Stochastic sliding mode control. J. Dyn. Syst. Meas. Control **138**(12), 124503(1–6) (2016)
34. Y. Wang, Z. Xia, Z. Jiang, G. Xie, A quasisliding mode variable structure control algorithm for discrete-time and time delay systems, in *IEEE Proceedings of Control Decision Conference*, 2011, pp. 107–110
35. Z. Gao, Integral sliding mode controller for an uncertain network control system with delay. Comput. Eng. Networking **277**(1), 31–38 (2013)
36. V. Goyal, V. Deolia, T. Sharma, Robust sliding mode control for nonlinear discrete-time delayed systems based on neural network. Intell. Control Autom. **6**(1), 75–83 (2015)
37. Y. Niu, D. Ho, Design of sliding mode control subject to packet losses. IEEE Trans. Autom. Control **55**(11), 2623–2628 (2010)
38. Y. Xiaojuan, L. Jinglin, BLDCM Fuzzy Sliding Mode Control based on network control system, in *International Conference on Advanced Computer Control*, 2010, pp. 332–335
39. Y. Yin, L. Xia, L. Song, M. Qian, Adaptive sliding mode control of networked control systems with variable time delay, in *International Conference on Electric and Electronics*, 2011, pp. 131–138
40. J. Zhang, J. Lam, Y. Xia, Output feedback sliding mode control under networked environment. Int. J. Syst. Sci. **44**(4), 750–759 (2013)
41. H. Li, H. Yang, F. Sun, Y. Xia, Sliding-mode predictive control of networked control systems under a multiple-packet transmission policy. IEEE Trans. Ind. Electron. **61**(11), 201–221 (2014)
42. A. Argha, L. Li, S.W. Su, H. Nguyen, Discrete-time sliding mode control for networked systems with random communication delays, in *Proceedings of American Control Conference*, 2015, pp. 6016–6021
43. H. Song, S. Chen, Y. Yam, Sliding mode control for discrete-time systems with Markovian packet dropouts. IEEE Trans. Cybern. **PP**(99), pp. 1–11 (2016)

Chapter 3
Discrete-Time Sliding Mode Controller for NCS with Deterministic Type Fractional Delay: A Switching Type Algorithm

Abstract In this chapter, a novel approach is presented for designing a discrete-time sliding mode controller. The effect of sensor to controller fractional delay and controller to actuator fractional delay in discrete-time domain is compensated through Thiran's approximation technique. The forward channel delay is compensated at the actuator side, while feedback channel delay is compensated at the sliding surface. An evolved sliding surface with delay compensation is used to derive SMC law. The stability condition for the closed-loop system with proposed controller is derived using Lyapunov function. The efficacy of the proposed algorithm is shown by simulation results and also validated by the experimental results considering DC servo system in networked environment having matched uncertainties.

Keywords Network delay · Thiran's approximation
Discrete-time sliding mode control · Stability

3.1 Network-Induced Fractional Delay Compensation with Thiran's Approximation

Figure 3.1 portrays the block diagram of NCS with network-induced time delay compensation scheme. It can be noticed that the state information as well as control information are transmitted to the controller and actuator through the network medium. During data transmission, the state information will experience sensor to controller delay, while the control information will suffer from controller to actuator delay. These delays are broadly defined as the amount of time required for the data packets to travel within the network. Thus, in order to avoid the degradation it is necessary to compensate these network delays at the controller end as well as at actuator end. Moreover, apart from these network delays, it is necessary to consider the system delays.

© Springer Nature Singapore Pte Ltd. 2018

D. H. Shah and A. Mehta, *Discrete-Time Sliding Mode Control for Networked Control System*, Studies in Systems, Decision and Control 132,
https://doi.org/10.1007/978-981-10-7536-0_3

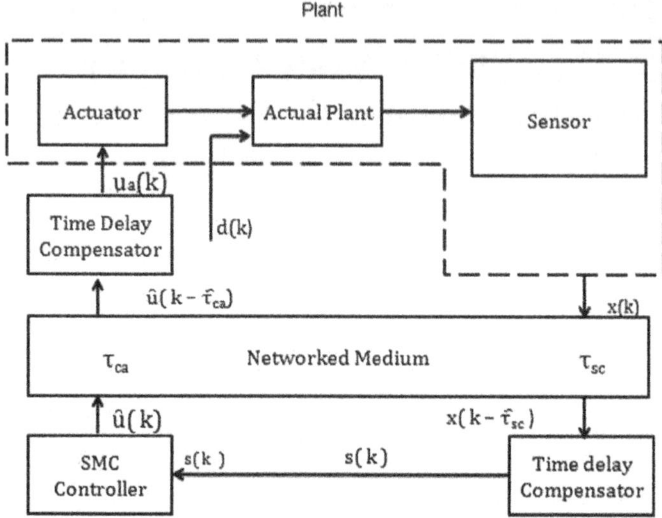

Fig. 3.1 Block diagram of NCS with time delay compensation

3.2   Problem Statement

Consider the linear time-invariant SISO system with network delay as:

$$\dot{x}(t) = Ax(t) + Bu(t - \tau) + Dd(t), \tag{3.1}$$

$$y(t) = Cx(t), \tag{3.2}$$

where $x \in R^n$ is system state vector, $u \in R^m$ is control input, $y \in R^p$ is system output, $A \in R^{n \times n}, B \in R^{n \times m}, C \in R^{p \times n}, D \in R^{n \times m}$ are the matrices of appropriate dimensions, $d(t)$ is the matched bounded disturbance with $|d(t)| \leq d_{max}$, and τ is the deterministic total networked-induced delay in continuous-time domain.

The discrete form of system (3.1) and (3.2) is:

$$x(k + 1) = Fx(k) + Gu(k - \tau') + d(k), \tag{3.3}$$

$$y(k) = Cx(k), \tag{3.4}$$

where τ' is the deterministic total fractional network-induced delay in discrete-time domain, $F = e^{Ah}$, $G = \int_0^h e^{At} B dt$, $d(k) = \int_0^h e^{At} Dd((k + 1)h - t)dt \in O(h)$. Since $|d(t)| \leq d_{max}$, it can be inferred that $d(k)$ is also bounded and $O(h)$ [1]. For simplicity, it is assumed that $d(k)$ is slowly varying and remains constant over the interval $kh \leq t \leq (k + 1)h$ [1].

The deterministic total fractional network-induced delay (τ') is denoted as

$$\tau' = \frac{\tau}{h},$$

where h is the sampling interval.

Remark 1 It is considered that network-induced fractional delay (τ') in discrete time has non-integer values so it is required to compensate the delay at each sampling instants.

Assumption 1 The total network-induced delay (τ) is deterministic in nature satisfying

$$\tau \prec h. \tag{3.5}$$

Remark 2 The above condition (3.5) indicates that the values of total fractional network-induced delay (τ') in discrete-time domain will be less than unity.

The total fractional network-induced delay (τ') is the combination of sensor to controller fractional delay (τ'_{sc}) and controller to actuator fractional delay (τ'_{ca}) which is given as

$$\tau' = \tau'_{sc} + \tau'_{ca}, \tag{3.6}$$

where $\tau'_{sc} = \frac{\tau_{sc}}{h}$ and $\tau'_{ca} = \frac{\tau_{ca}}{h}$.

Assumption 2 The disturbance $d(k)$ is bounded by upper and lower bounds as:

$$d_l \leq d(k) \leq d_u, \tag{3.7}$$

where d_l and d_u denote lower and upper bounds of disturbance.

Remark 3 Without loss of generality, the sensor processing delay (τ_{sp}), controller computational delay (τ_{cp}) and actuator processing delay (τ_{ap}) are neglected as their values are negligible compared to network-induced delay (τ).

Now, we are ready to define the problem statement with above assumptions and conditions.

Problem Statement To design robust discrete-time sliding mode controller for the system (3.3, 3.4) in the presence of deterministic fractional network delays τ'_{sc} and τ'_{ca} under the Assumptions (1) and (2).

The sliding mode controller design involves the sliding surface design and the control law that computes the control sequences and steers the states towards the surface.

The next section proposes the design of sliding surface that compensates the effect of fractional delay occurring from sensor to controller.

3.3 Sliding Surface Design for Deterministic Type Network-Induced Delay

There are two widely used approaches, namely Tustin approximation and bilinear transformation for time delay compensation in discrete-time domain. However, the limitation of both the approaches is that they cannot approximate fractional delay which is of main concern here [2–4]. The Thiran approximation [5] technique approximates the non-integer types of delays in discrete-time domain. Thiran has proposed the time delay approximation algorithm for maximally flat group of fractional delays occurring in signal processing applications. Hence, it is proper candidate for fractional delay compensation for discrete-time SMC design.

The fractional delay in discrete time can be approximated by Thiran's approximation as under:

$$z^{-\nu} = \Sigma_{k=0}^{l}(-1)^{k}\binom{l}{k}\Pi_{i=0}^{l}\frac{2\tau_{sc}' + i}{2\tau_{sc}' + k + i}z^{-k}, \tag{3.8}$$

where l indicates the order of approximation, $\nu = \frac{\delta}{h}$ indicates the fractional part of delay, δ is the delay occurring during signal transmission, and h is the sampling interval.

The order of approximation is given by:

$$l = ceil(\nu), \tag{3.9}$$

where $ceil$ operator rounds the nearest positive integer greater than or equal to ν. Next, the sliding surface using above approximation is proposed as Lemma 1 given below.

Lemma 1 *The compensated sliding variable $s(k)$ for the given system (3.3, 3.4) with sensor to controller network-induced fractional delay (τ_{sc}') satisfying condition (3.5) and under the Assumptions (1) and (2) is given as:*

$$s(k) = C_{s}x(k) - \alpha C_{s}(x(k-1)), \tag{3.10}$$

where
$\alpha = \frac{\tau_{sc}'}{\tau_{sc}'+1}$ and C_{s} is the sliding gain.

Proof The sliding variable with the delayed state vector at the receiving end of controller is given by:

$$s(k) = C_{s}x(k - \tau_{sc}'), \tag{3.11}$$

where τ_{sc}' is the sensor to controller fractional delay. The sliding gain C_{s} is calculated using discrete LQR method through proper selection of Q and R matrices [6].

Applying z-transform to Eq. (3.11), we get

$$s(z) = C_s x(z) z^{-\tau'_{sc}}, \qquad (3.12)$$

where $\tau'_{sc} = \frac{\tau_{sc}}{h}$.

$z^{-\tau'_{sc}}$ can be approximated as [5]

$$z^{-\tau'_{sc}} = \Sigma_{k=0}^{1}(-1)^k \binom{l}{k} \Pi_{i=0}^{1} \frac{2\tau'_{sc}+i}{2\tau'_{sc}+k+i} z^{-k}. \qquad (3.13)$$

The above Eq. (3.13) can be further expanded as

$$z^{-\tau'_{sc}} = [(-1)^0 \binom{1}{0} \left\{ \frac{2\tau'_{sc}}{2\tau'_{sc}} * \frac{2\tau'_{sc}+1}{2\tau'_{sc}+1} \right\} z^0 + (-1)^1 \binom{1}{1} \qquad (3.14)$$
$$\left\{ \frac{2\tau'_{sc}}{2\tau'_{sc}+1} * \frac{2\tau'_{sc}+1}{2\tau'_{sc}+2} \right\} z^{-1}].$$

On simplifying, we get

$$z^{-\tau'_{sc}} = 1 - \alpha z^{-1}, \qquad (3.15)$$

where $\alpha = \frac{\tau'_{sc}}{\tau'_{sc}+1}$.

To show the effect of Thiran's approximation, the step response of the compensated system with the system having no delay is shown in Figs. 3.2 and 3.3, respectively. Figure 3.2 shows the step response for $\tau'_{sc} = 1$, and Fig. 3.3 shows the step response for $\tau'_{sc} = 0.5$. In both the cases, it can be noticed that Thiran's approximation approximates the fractional delay accurately and at each sampling instants the effect of fractional delay is nullified.

Thus, substituting Eq. (3.15) into (3.12),

$$s(z) = C_s x(z)[1 - \alpha z^{-1}]. \qquad (3.16)$$

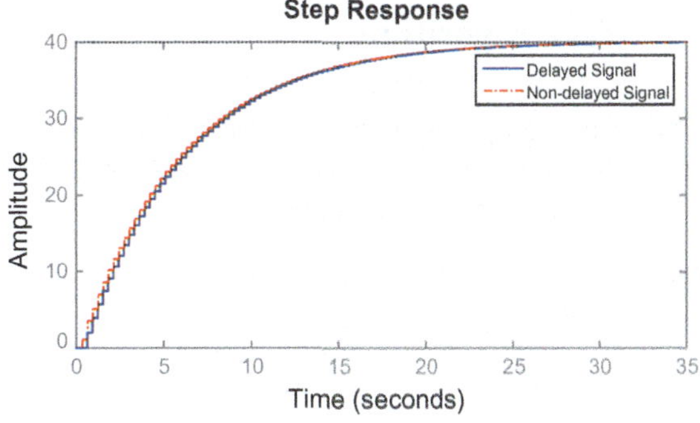

Fig. 3.2 Step response of Thiran's approximation with $\tau'_{sc} = 1$

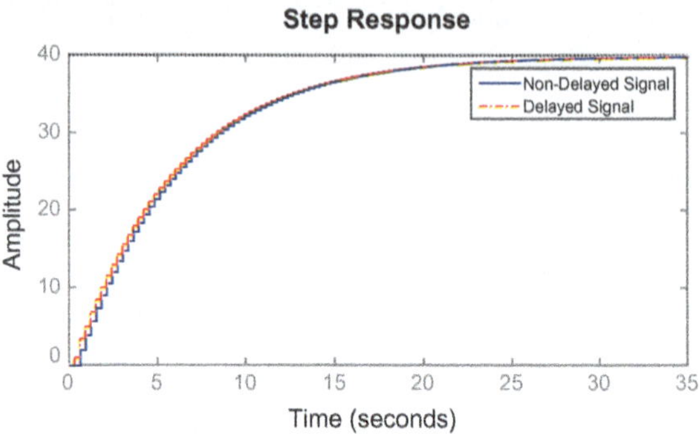

Fig. 3.3 Step response of Thiran's approximation with $\tau'_{sc} = 0.5$

Further expanding, we may get

$$s(z) = C_s x(z) - \alpha C_s z^{-1} x(z). \tag{3.17}$$

Applying inverse z-transform, we may have

$$s(k) = C_s x(k) - \alpha C_s s x(k-1). \tag{3.18}$$

This completes the *proof*.

From Eq. (3.18), it is inferred that the network-induced fractional delay from sensor to controller can be compensated in the sliding surface $s(k)$ at each sampling instant h using the current and immediate past sample information and parameter α.

Now, we are ready to design a discrete-time sliding mode control law using the proposed sliding surface (3.18).

3.4   Design of Discrete-Time Sliding Mode Control for NCS Using Thiran's Delay Approximation: A Switching Type Algorithm

This section proposes switching type control law based on Gao's reaching law [7] and sliding surface (3.18). The Gao's reaching law provides the faster convergence within the specified quasi-sliding mode band.

Theorem 3.1 *The discrete-time sliding mode controller for system* (3.3, 3.4) *in the presence of deterministic fractional delays and matched uncertainty* $d(k)$ *is given as*

$$u(k) = -(C_s G)^{-1}[M x(k) - N x(k) - (1 - qh)(s(k)) + \varepsilon h sgn(s(k))] - d(k). \tag{3.19}$$

where

$M = (C_s F)$ *and* $N = \alpha C_s$.

Proof Let us consider the Gao's reaching law [7] as:

$$s[(k + 1)] = (1 - qh)s(k) - \varepsilon h sgn(s(k)), \qquad (3.20)$$

where
$q, \varepsilon > 0, 0 \prec (1 - qh) \prec 1$, sgn is the signum function, h represents the sampling interval, and $s(k)$ is the sliding surface proposed in (3.18).

The reaching law in Eq. (3.20) states that the state vector always moves towards the quasi-sliding mode band given as:

$$s(k) \leq \frac{\varepsilon h}{2 - qh}. \qquad (3.21)$$

Substituting Eq. (3.18) in Eq. (3.20), we may get

$$C_s x(k + 1) - \alpha C_s x(k) = (1 - qh)s(k) - \varepsilon h sgn(s(k)).$$

Remark 4 It is noticed from Eq. (3.18) that the sensor to controller fractional delay is compensated at the sliding surface, while controller to actuator fractional delay is compensated at the actuator side. So, the effect of delay will not be observed at the controller side for computing the control signal as well at actuator side. Thus without loss of generality, the control signal in Eq. (3.3) is given as

$$u(k - \tau') = u(k) \qquad (3.22)$$

Substituting $x(k + 1)$ from Eq. (3.3), we get

$$C_s[Fx(k) + Gu(k) + d(k)] - \alpha C_s x(k) = (1 - qh)s(k) - \varepsilon h sgn(s(k)). \qquad (3.23)$$

Further, simplifying we may write above Eq. (3.23) as

$$C_s Fx(k) + C_s Gu(k) + C_s d(k) - \alpha C_s x(k) = (1 - qh)s(k) - \varepsilon h sgn(s(k)). \qquad (3.24)$$

Further, above Eq. (3.24) can be expressed as a control law

$$u(k) = -(C_s G)^{-1}[Mx(k) - Nx(k) + (1 - qh)(s(k)) - \varepsilon h sgn(s(k))] - d(k). \qquad (3.25)$$

where

$M = (C_s F)$ and $N = \alpha C_s$.
This completes the *proof*.

The closed-loop stability is derived using compensated sliding surface (3.18) and control law proposed in Eq. (3.25) such that the system states remain within specified band (3.21) for a finite interval of time.

3.5 Stability Analysis

Theorem 3.2 *The state trajectories of the closed-loop system (3.3, 3.4) with network delay (τ') and matched uncertainty $d(k)$ with the controller (3.25) drive towards the sliding surface (3.18) and maintain on it for any $q, \varepsilon, \beta > 0, 0 \prec 1 - qh \prec 1$ and $1 - qh \prec \varepsilon$ provided the following condition holds true:*

$$0 \preceq \Phi \prec s^T(k)s(k). \tag{3.26}$$

where

$$\Phi = [(1 - qh)s(k) - \varepsilon hsgn(s(k))]^T * [(1 - qh)s(k) - \varepsilon hsgn(s(k))]$$

Proof Let us consider sliding surface (3.18) as

$$s(k) = C_s x(k) - \alpha C_s x(k - 1). \tag{3.27}$$

Selecting the Lyapunov function as

$$V_s(k) = s^T(k)s(k). \tag{3.28}$$

Writing forward difference of the above Eq. (3.28),

$$\Delta V_s(k) = s^T(k + 1)s(k + 1) - s^T(k)s(k). \tag{3.29}$$

Substituting the value of $s(k + 1)$ using Eq. (3.27), we get

$$\Delta V_s(k) = [C_s x(k + 1) - \alpha C_s x(k)]^T [C_s x(k + 1) \\ -\alpha C_s x(k)] - s^T(k)s(k). \tag{3.30}$$

Substituting the value of $x(k+1)$,

$$\Delta V_s(k) = [C_s[Fx(k) + G(u(k) + d(k))] - \alpha C_s x(k)]^T$$
$$[C_s Fx(k) + G(u(k) + d(k))] - \alpha C_s x(k)] - s^T(k)s(k). \tag{3.31}$$

Substituting the value of $u(k)$ from Eq. (3.25) and further solving it, we have

$$\Delta V_s(k) = [(1 - qh)s(k) - \varepsilon h sgn(s(k))]^T * [(1 - qh) \tag{3.32}$$
$$s(k) - \varepsilon h sgn(s(k))] - s^T(k)s(k).$$

Denoting,

$$\Phi = [(1 - qh)s(k) - \varepsilon h sgn(s(k))]^T * [(1 - qh)s(k) - \varepsilon h sgn(s(k))]$$

Then, we have
$$\Delta V_s(k) = \Phi - s^T(k)s(k). \tag{3.33}$$

The term Φ can be tuned close to zero by appropriately selecting the parameter q and ε. If Φ is close to zero, then $s^T(k)s(k)$ will be larger than Φ. Thus, for any small parameter β, we have $\Phi - s^T(k)s(k) \prec \beta s^T(k)s(k)$.

Thus, by tuning the parameter q and ε, we have $\Delta V_s(k) \prec \beta s^T(k)s(k)$ which guarantees the convergence of $\Delta V_s(k)$ and implies that any trajectory of the system (3.3, 3.4) will be driven onto the sliding surface and maintain on it.
This completes the *proof*.

The control signal $u(k)$ computed in (3.25) using compensated sliding surface (3.10) will also experience controller to actuator fractional delay (τ'_{ca}) which results in the delayed control signal $u(k - \tau'_{ca})$. So, in order to avoid the degradation of the plant response again the time delay is compensated from controller to actuator. The compensated control signal at the actuator end can be represented as:

$$u_a(k) = u(k) - \alpha' u(k - 1), \tag{3.34}$$

where
$$\alpha' = \frac{\tau'_{ca}}{1 + \tau'_{ca}}.$$

It can be noticed from above Eq. (3.34) that the compensated control signal $u_a(k)$ depends on difference of the present control signal that is available from network as well as past control signal which is multiplied over the parameter α' approximated through Thiran's approximation. Thus, the effect of controller to actuator fractional delay is compensated at actuator side which is further applied to the plant.

3.6 Simulation and Experimental Results

3.6.1 System Description

In this section, Quanser Qnet 2.0 brushed DC motor setup is explained in detail on which the simulation as well as experimental results are carried out using proposed control law. The performance of the brushless DC motor is tested under different networked delays as well external disturbances to prove the robustness of the proposed algorithm. The results obtained with proposed algorithm are compared with the conventional SMC algorithm.

Figure 3.4 shows the block diagram of Quanser made Qnet 2.0 brushed DC motor setup used for the simulation as well as experimental purpose. The Qnet DC motor provides an integrated amplifier and a communication interface with the NI ELVIS II (+) for the amplifier command and encoder port [8]. The NI ELVIS II (+) is interfaced to the PC via USB link to the Qnet DC motor setup as shown in Fig. 3.5. The NI ELVIS II (+) block reads the angular encoder as an input and commands the power

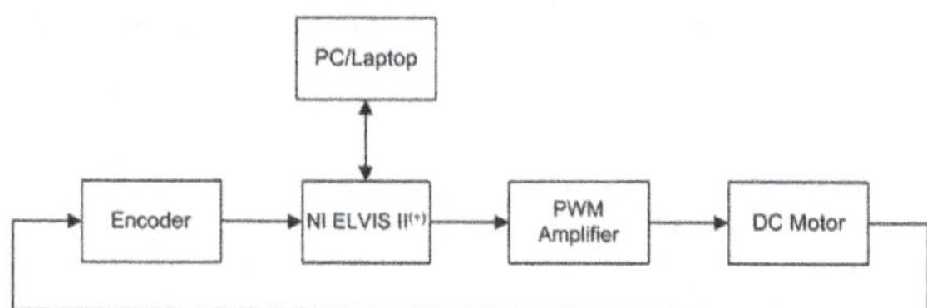

Fig. 3.4 Block diagram of Qnet DC servo motor components

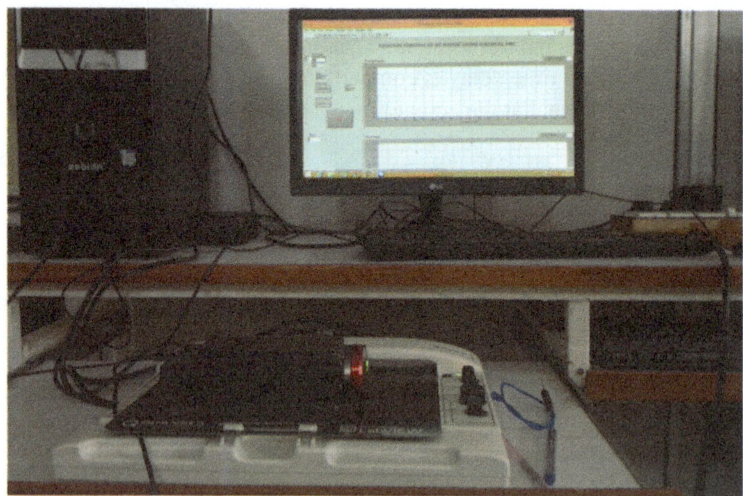

Fig. 3.5 Experimental setup of Quanser DC servo motor

amplifier which acts as driver for the motor. The various network delays are generated through software blocks.

The detailed mathematical model along with the system parameters of the DC motor is given as [9]:

$$\frac{\theta(s)}{V_m(s)} = \frac{K_m}{J_m R_m s^2 + K_m^2 s},$$

(3.35)

where

$\theta(s)$ = output from the system (position),
V_m = input to the system,
J_m = rotor inertia = 4×10^{-6} kgm^2,
R_m = terminal resistance = $8.4\,\Omega$,
K_m = motor back emf constant = 0.042 V/(rad/s).

Substituting these parameters, the state space model of the above system (3.35) is given as

$$\dot{x}(t) = Ax(t) + Bu(t - \tau) + Dd(t),$$

(3.36)

$$y(t) = Cx(t),$$

(3.37)

where

$$A = \begin{bmatrix} -201 & 0 \\ 1 & 0 \end{bmatrix}, \ B = \begin{bmatrix} 1 \\ 0 \end{bmatrix},$$

$$C = \begin{bmatrix} 0 & 1 \end{bmatrix}, \ D = \begin{bmatrix} 1 \\ 1 \end{bmatrix}, \ d(t) = 0.2 \sin (0.086t).$$

Discretizing the system at sampling interval $h = 30$ ms, we get

$$x(k + 1) = Fx(k) + Gu(k - \tau') + d(k),$$

(3.38)

$$y(k) = Cx(k),$$

(3.39)

where

$$F = \begin{bmatrix} 0.001836 & 0 \\ 0.004753 & 1 \end{bmatrix}, \ G = \begin{bmatrix} 0.004753 \\ -0.0001242 \end{bmatrix},$$

$$C = \begin{bmatrix} 0 & 1 \end{bmatrix}.$$

3.6.2 Discussion of Simulation and Experimental Results

In this section, the simulation and experimental results of position control of DC motor are thoroughly discussed in the presence of deterministic network delays. Figures 3.6, 3.7, 3.8, 3.9 and 3.10 show the simulation and experimental responses of system for trajectory tracking, compensated sliding variable and control efforts under different network delays. To show the robustness properties, slow time-varying disturbance signal is applied through the input channel. The total networked-induced

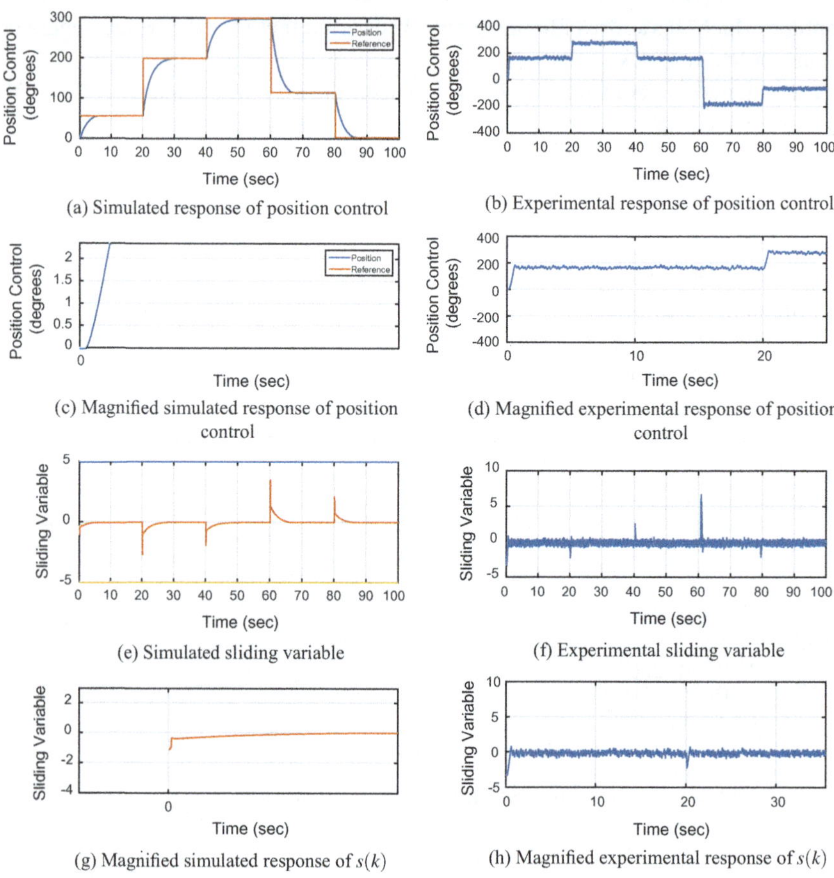

(a) Simulated response of position control

(b) Experimental response of position control

(c) Magnified simulated response of position control

(d) Magnified experimental response of position control

(e) Simulated sliding variable

(f) Experimental sliding variable

(g) Magnified simulated response of $s(k)$

(h) Magnified experimental response of $s(k)$

Fig. 3.6 Simulation and experimental results for position tracking and compensated sliding surface for $\tau = 12.8\,\text{ms}$

delay with a range of 12.8–28 ms is generated through network block for which the effect of delay is compensated satisfying condition (3.5). The position of DC motor is considered as the reference input. The sliding gain is computed using discrete LQR method which comes out to be $C_s = [2.5156\ 31.6288]$ with $Q = diag(1000, 1000)$ and $R = 1$, while the quasi-sliding band (3.21) comes out to be $|s(k)| \leq -5$ to 5 with tuning parameters $q = 30$ and $\varepsilon = 2000$.

Figures 3.6a and 3.7d show the simulation and experimental results of position control of DC motor plant for total network delay of $\tau = 12.8\,\text{ms}$ with $\tau_{sc} = 6.4\,\text{ms}$ and $\tau_{ca} = 6.4\,\text{ms}$. The fractional part of total network delay is obtained as $\tau' = 0.426$, $\tau'_{sc} = 0.213$ and $\tau'_{ca} = 0.213$ for $h = 30\,\text{ms}$. The trajectory response of the system in case of simulation and experimental results is shown in Fig. 3.6a, b, respectively. In both cases, the output tracks the reference trajectory in the presence of specified network delay. In order to show the exact effect of time delay compensation

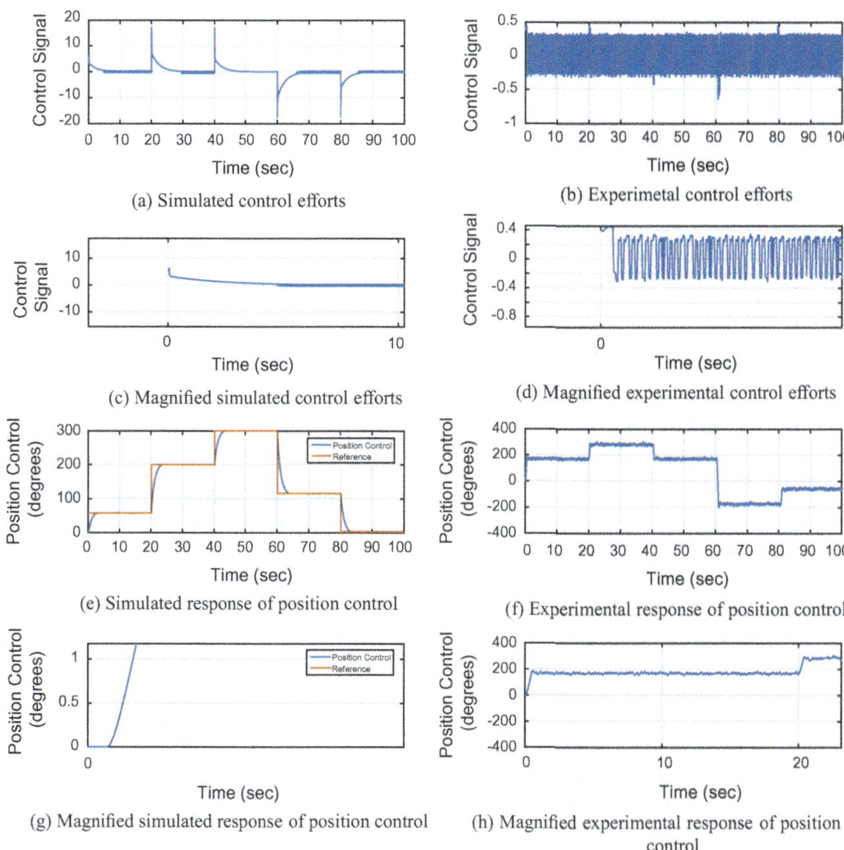

(a) Simulated control efforts

(b) Experimetal control efforts

(c) Magnified simulated control efforts

(d) Magnified experimental control efforts

(e) Simulated response of position control

(f) Experimental response of position control

(g) Magnified simulated response of position control

(h) Magnified experimental response of position control

Fig. 3.7 **a–d** Simulation and experimental result of control efforts for $\tau = 12.8$ ms and **e–h** simulation and experimental result of position control for $\tau = 24$ ms

at the output, results are magnified as shown in Figs. 3.6c, d. It can be noticed that the effect of fractional time delay from sensor to controller is compensated as the output tracks the trajectory at 6.4 ms. The same effect of time delay compensation from sensor to controller can be observed in sliding surface as shown in Fig. 3.6e, f as well as control signal as shown in Fig. 3.7a, b. Observing the magnified results of Fig. 3.6c, d, g, h, it can be noticed that both the sliding surface and control signal are computed at first sampling instant even in the presence of sensor to controller delay. Thus, the effects of fractional delay from sensor to controller at sliding surface and control signal are compensated and remain within the specified sliding band (3.21). The proposed algorithm was further extended for higher values of τ. Figures 3.7e and 3.8h show the simulation and experimental results of position control of DC motor for total networked delay of $\tau = 24$ ms with $\tau_{sc} = 12$ ms and $\tau_{ca} = 12$ ms. The fractional part of total network delay is computed as $\tau' = 0.8$, $\tau'_{sc} = 0.4$ and

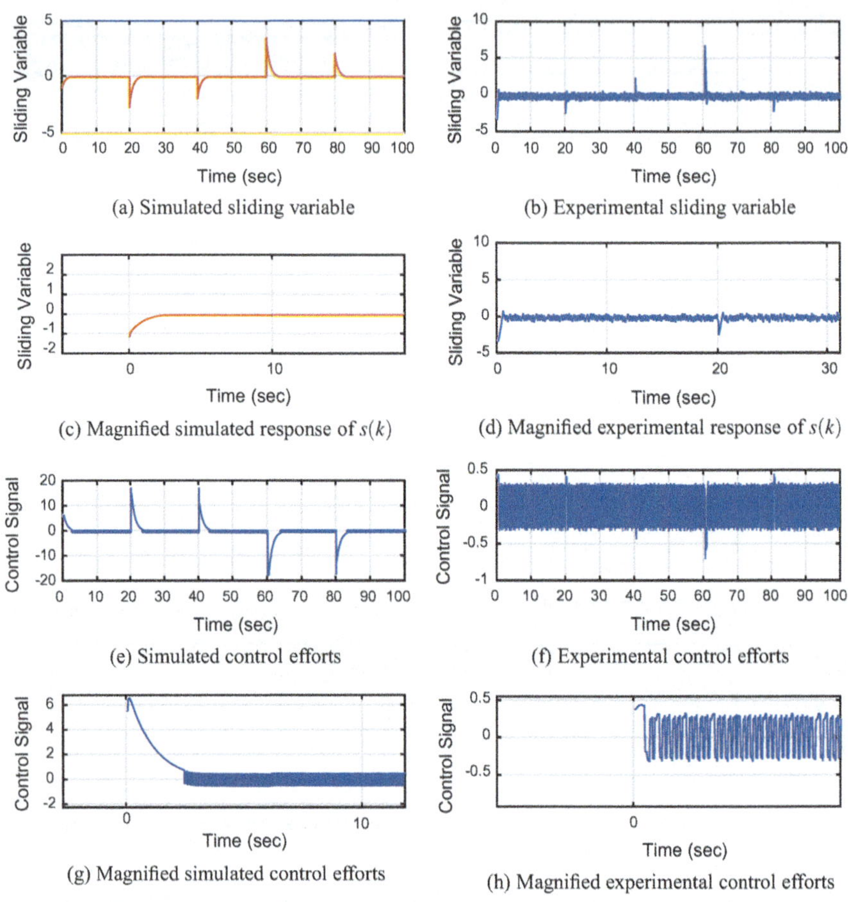

Fig. 3.8 Simulation and experimental result of sliding variable and control signal for $\tau = 24$ ms

$\tau'_{ca} = 0.4$ for $h = 30$ ms. The simulated and experimental trajectory response of the system are shown in Fig. 3.7e, f respectively. Observing the results, it can be noticed the output tracks the reference signal in the specified networked delay. In order to show the effect of delay compensation, the output results are magnified as shown in Fig. 3.7g and h, respectively. It can be noticed that the effect of fractional delay from sensor to controller is nullified as the output tracks the reference trajectory at $t = 12$ ms. The similar effect of time delay compensation can be observed in sliding surface as well as control efforts signal as shown in Fig. 3.8a–h. Observing the simulated and experimental magnified results of sliding surface (Fig. 3.8c, d) as well as control signal (Fig. 3.8g, h), it can be noticed that in both the cases the sliding surface and control signal are computed at first sampling instant. Thus, the fractional delay from sensor to controller is compensated and remains within the specified sliding band (3.21). Figures 3.9a and 3.10c show the simulation and experimental

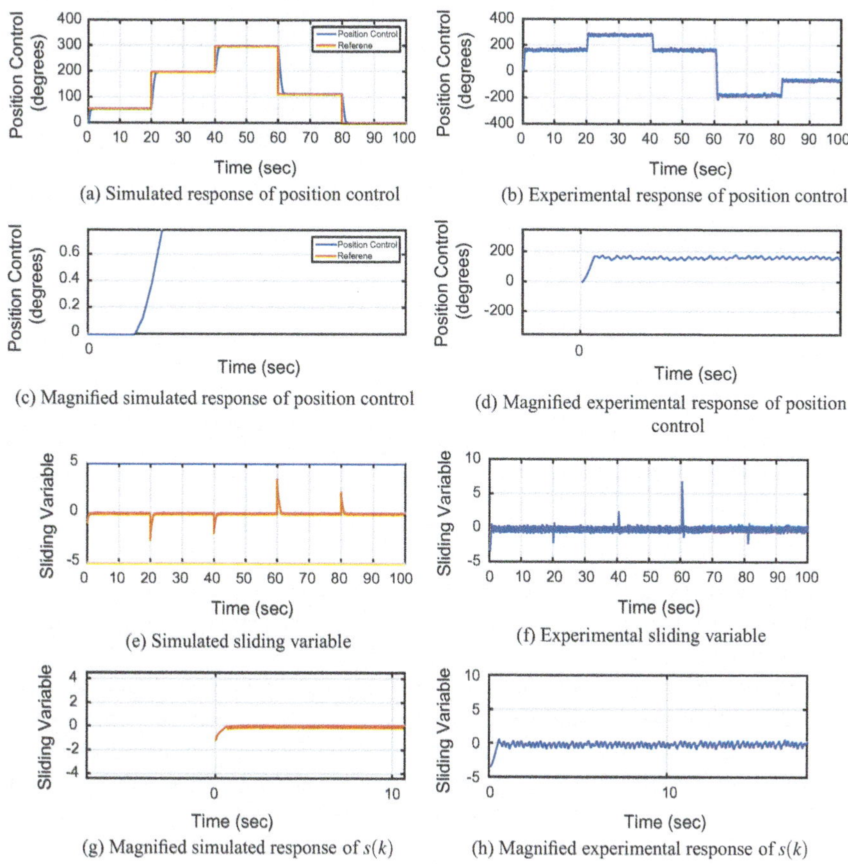

(a) Simulated response of position control

(b) Experimental response of position control

(c) Magnified simulated response of position control

(d) Magnified experimental response of position control

(e) Simulated sliding variable

(f) Experimental sliding variable

(g) Magnified simulated response of $s(k)$

(h) Magnified experimental response of $s(k)$

Fig. 3.9 Simulation and experimental result for tracking position and sliding surface for $\tau = 28$ ms

results of position control of DC motor for total networked delay of $\tau = 28$ ms with $\tau_{sc} = 14$ ms and $\tau_{ca} = 14$ ms. The fractional part of total networked delay for $h = 30$ ms is obtained as $\tau' = 0.933$, $\tau'_{sc} = 0.466$ and $\tau'_{ca} = 0.466$, respectively. The simulation and experimental results with magnified response of reference trajectory are shown in Fig. 3.9a–d, respectively. Observing the results, it can be concluded that the output tracks the reference trajectory at $t = 14$ ms for the specified networked delay. Thus, the effect of fractional delay from sensor to controller is nullified at the output as shown in Fig. 3.9c, d. The similar effect of time delay compensation will be observed in sliding surface and control signal results as shown in Figs. 3.9e and 3.10c. Observing the results, it can be noticed that in simulation as well as experimental case the sliding surface and the control signal are computed from first sampling instant. Thus, the effect of fractional delay from sensor to controller is compensated at sliding surface as well as at control signal. Apart from delay compensation, the

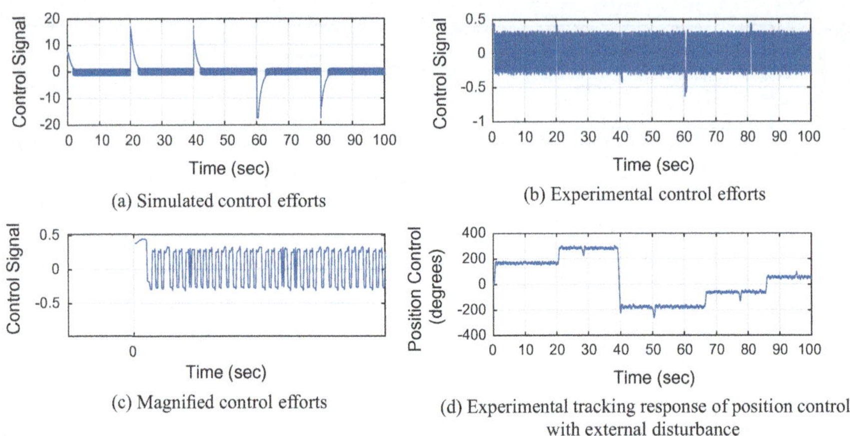

Fig. 3.10 Simulation and experimental result of control efforts for $\tau = 28$ ms with external disturbances

position of motor was also controlled by applying the external disturbances through rotating the wheel in forward and reverse directions. The situation of motor under external disturbances is shown in Fig. 3.10d.

Thus, from all the results it can be concluded that the proposed algorithm works efficiently with network delay range of 12.8 ms $\leq \tau \leq 28$ ms in experimental as well as in simulated environment. The proposed controller compensates the network time delay for $q = 30$ and $\varepsilon = 2000$ satisfying (3.5) and shows the stable response satisfying condition (3.26) in the presence of matched uncertainty.

3.6.2.1 Comparison of Proposed Algorithm with Conventional Sliding Mode Control

In this section, the experimental results of proposed algorithm are compared with conventional sliding mode control. The results of tracking response, control signal and sliding variable for total networked delay of $\tau = 12.8$ ms are shown in Fig. 3.11a–f. From the comparative results, it can be noticed that the conventional sliding mode control becomes unstable for a small delay of $\tau_{sc} = 6.4$ ms. Thus, the Thiran approximation proved to be an efficient method in discrete-time sliding mode control. Simulation and experimental results for different networked delays range are summarized in Table 3.1. The comparison of discrete-time sliding mode control with time delay approximation and conventional sliding mode control is shown in Table 3.2.

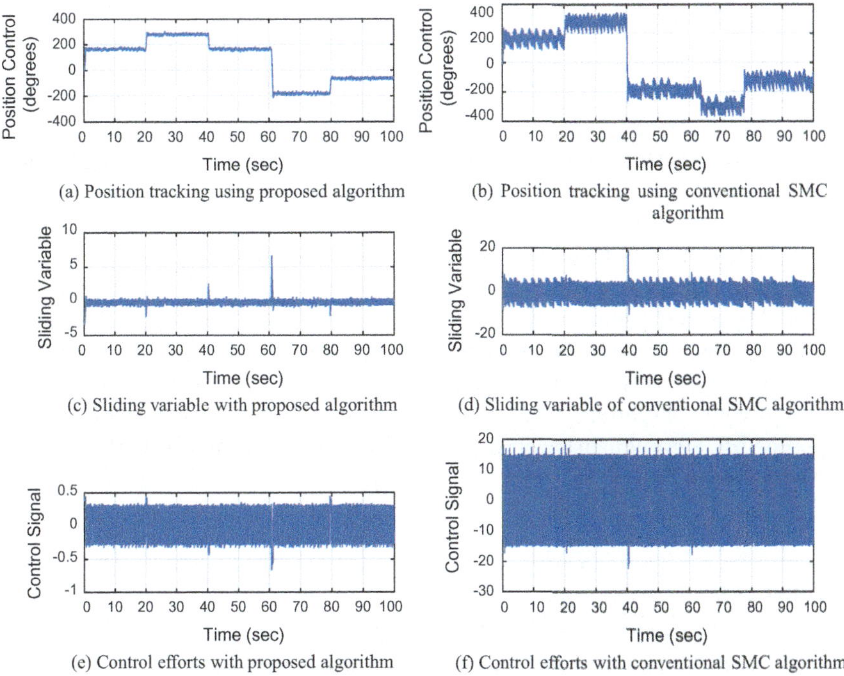

(a) Position tracking using proposed algorithm

(b) Position tracking using conventional SMC algorithm

(c) Sliding variable with proposed algorithm

(d) Sliding variable of conventional SMC algorithm

(e) Control efforts with proposed algorithm

(f) Control efforts with conventional SMC algorithm

Fig. 3.11 Comparative responses with proposed algorithm and conventional sliding mode control for DC motor position control for $\tau = 12.8\,\text{ms}$

Table 3.1 Simulation and experimental results with different networked-induced delays

Network delays (τ)	Simulations and experimental results	
	Chattering	Response
12.8 ms	Within QSMB	Satisfactory
24 ms	Within QSMB	Satisfactory
28 ms	Within QSMB	Satisfactory

Table 3.2 Comparison of proposed algorithm with conventional SMC

Algorithm	Comparative results			
	τ (ms)	T_s	Chattering	Response
Conventional SMC	12.8	Undefined	High	Unstable
Proposed method	12.8	1 s	Within QSMB	Stable

3.7 Conclusion

In this chapter, we explored Thiran's approximation technique for fractional delay compensation in discrete-time domain. The effect of network-induced fractional delay generated due to the communication medium is compensated in sliding surface. The sliding surface is designed in such a manner that the system states slide along the predetermined surface according to network delay. A switching type discrete-time sliding mode controller is designed which computes the control actions in the presence of network delay and matched uncertainty. The stability of the closed-loop NCS is assured by using Lyapunov approach. The effectiveness of the proposed algorithm is tested in simulation and experimental environment on brushless DC motor with deterministic type networked delay and matched uncertainty. The results are also compared with conventional sliding mode. The comparative results endow that the proposed SMC algorithm in the presence of fractional delay compensated by Thiran's approximation technique performs well in the presence of matched uncertainties.

References

1. A. Mehta, B. Bandyopadhyay, Multirate output feedback based stochastic sliding mode control. J. Dyn. Syst. Meas. Control **138**(12), 124503(1–6) (2016)
2. D. Shah, A. J. Mehta, Design of robust controller for networked control system, *Proceedings of IEEE International Conference on Computer, Communication and Control Technology* (Sept. 2014), pp. 385–390
3. D. Shah, A. Mehta, Discrete-time sliding mode controller subject to real-time fractional delays and packet losses for networked control system. Int. J. Control, Autom. Syst. (IJCAS), **15**(6), 2690–2703 (Dec. 2017)
4. D. Shah, A. Mehta, Fractional delay compensated discrete-time SMC for networked control system. Digital Commun. Networks (DCN), Elsevier, **2**(3), 385–390 (Dec. 2016)
5. J. Thiran, Recursive digital filters with maximally flat group delay. IEEE Trans. Circ. Theory **18**(6), 659–664 (1971)
6. S. Eduardo, *Mathematical Control Theory: Deterministic Finite Dimensional Systems*, 2nd ed. (Springer, Berlin, 1998), ISBN 0-387-98489-5
7. W. Gao, Y. Wang, A. Homaifa, Discrete-time variable structure control systems. IEEE Trans. Ind. Electron. **42**(2), 117–122 (1995)
8. D. Shah, A. Mehta, Discrete-time sliding mode control using Thiran's delay approximation for networked control system, in *43rd Annual Conference on Industrial Electronics (IECON-17)*, pp. 3025–3031 (Nov. 2017), ISBN 978-1-5386-1126-5
9. K. Astrom, J. Apkarian, P. Karam, M. Levis, J. Falcon, *Student Workbook: QNET DC Motor Control Trainer for NI ELVIS* (Quanser, 2015)
10. W. Yang, L. Fan, J. Luo, Design of discrete time sliding mode observer in networked control system, in *2010 Chinese Control and Decision Conference*, (July 2010) pp. 1884–1887

Chapter 4
Discrete-Time Sliding Mode Controller for NCS with Deterministic Fractional Delay: A Non-switching Type Algorithm

Abstract In this chapter, the design of discrete-time sliding mode controller using Thiran's delay approximation is extended for non-switching type algorithm. The effect of sensor to controller delay is compensated using Thiran's delay approximation technique in the sliding surface. Further, Lyapunov approach is used to determine the stability of closed-loop NCSs with the proposed controller. The efficacy of the control methodology is endowed by simulation and experimental results in the presence of networked delay. The performance of the proposed control algorithm is further validated in the presence of real-time networks such as CAN and Switched Ethernet using true time simulator.

Keywords Deterministic network delay · Thiran's approximation
True time simulator · CAN · Switched Ethernet
Discrete-time sliding mode control · Stability

4.1 Network-Induced Fractional Delay Compensation

Figure 4.1 portrays the block diagram of NCS with network-induced time delay compensation scheme. It can be noticed that the state information as well as control information is transmitted to the controller and actuator through the network medium. During data transmission, the state information will experience sensor to controller delay, while the control information will suffer from controller to actuator delay. These delays are broadly defined as the amount of time required for the data packets to travel within the network [1, 2]. Thus, in order to avoid the degradation, it is necessary to compensate these network delays at the controller end as well as at actuator end. Moreover, apart from these network delays, it is necessary to consider the system delays.

© Springer Nature Singapore Pte Ltd. 2018

D. H. Shah and A. Mehta, *Discrete-Time Sliding Mode Control for Networked Control System*, Studies in Systems, Decision and Control 132,
https://doi.org/10.1007/978-981-10-7536-0_4

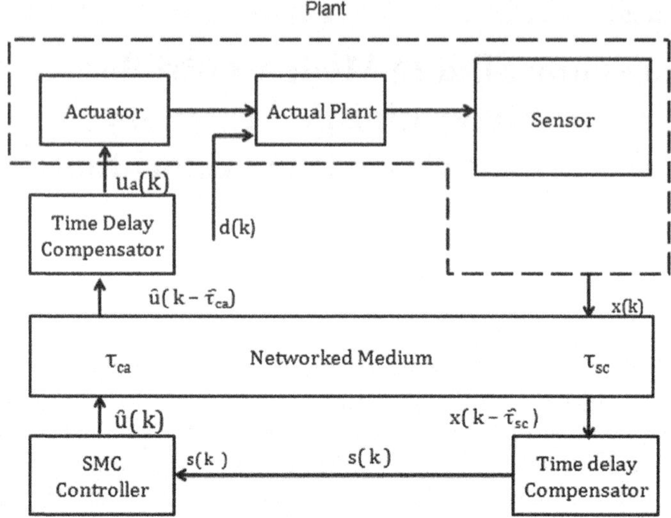

Fig. 4.1 Block diagram of NCS with time delay compensation

4.2 Problem Statement

Consider the linear time invariant SISO system with network delay as:

$$\dot{x}(t) = Ax(t) + Bu(t - \tau) + Dd(t), \tag{4.1}$$

$$y(t) = Cx(t), \tag{4.2}$$

where $x \in R^n$ is system state vector, $u \in R^m$ is control input, $y \in R^p$ is system output, $A \in R^{n \times n}$, $B \in R^{n \times m}$, $C \in R^{p \times n}$, $D \in R^{n \times m}$ are the matrices of appropriate dimensions, $d(t)$ is the matched bounded disturbance with $|d(t)| \leq d_{max}$, and τ is the deterministic total networked induced delay in continuous-time domain.

The discrete form of system (4.1) and (4.2) is:

$$x(k + 1) = Fx(k) + Gu(k - \tau') + d(k), \tag{4.3}$$

$$y(k) = Cx(k), \tag{4.4}$$

where τ' is the deterministic total fractional network-induced delay in discrete-time domain, $F = e^{Ah}$, $G = \int_0^h e^{At}Bdt$, $d(k) = \int_0^h e^{At}Dd((k + 1)h - t)dt \in O(h)$. Since $|d(t)| \leq d_{max}$, it can be inferred that $d(k)$ is also bounded and $O(h)$ [3]. For simplicity, it is assumed that $d(k)$ is slowly varying and remain constant over the interval $kh \leq t \leq (k + 1)h$ [3].

The deterministic total fractional network-induced delay (τ') is denoted as,

$$\tau' = \frac{\tau}{h},$$

where h is the sampling interval.

Remark 5 It is considered that network-induced fractional delay (τ') in discrete-time has non-integer values so it is required to compensate the delay at each sampling instants.

Assumption 3 The total network-induced delay (τ) is deterministic in nature satisfying,

$$\tau \prec h, \tag{4.5}$$

Remark 6 The above condition (4.5) indicates that the values of total fractional network-induced delay (τ') in discrete-time domain will be less than unity.

The total fractional network-induced delay (τ') is the combination of sensor to controller fractional delay (τ'_{sc}) and controller to actuator fractional delay (τ'_{ca}) which is given as,

$$\tau' = \tau'_{sc} + \tau'_{ca}, \tag{4.6}$$

where $\tau'_{sc} = \frac{\tau_{sc}}{h}$ and $\tau'_{ca} = \frac{\tau_{ca}}{h}$.

Assumption 4 The disturbance $d(k)$ is bounded by upper and lower bounds as:

$$d_l \le d(k) \le d_u, \tag{4.7}$$

where d_l and d_u denotes lower and upper bounds of disturbance.

Remark 7 Without loss of generality, the sensor processing delay (τ_{sp}), controller computational delay (τ_{cp}) and actuator processing delay (τ_{ap}) are neglected as their values are negligible compared to network-induced delay (τ).
Now, we are ready to define the problem statement with above assumptions and conditions.

Problem Statement: To design robust discrete-time sliding mode controller for the system (4.3, 4.4) in the presence of deterministic fractional network delays τ'_{sc} and τ'_{ca} under the Assumptions (3) and (4).

The sliding mode controller design involves the sliding surface design and the control law that computes the control sequences and steers the states towards the surface.

The next section proposes the design of sliding surface that compensates the effect of fractional delay occurring from sensor to controller.

4.3 Sliding Surface Design

There are two widely used approaches namely Tustin approximation and bilinear transformation for time delay compensation in discrete-time domain. However, the limitation of both the approaches is that they cannot approximate fractional delay which is of main concern here. The Thiran approximation [4] technique approximates the non-integer types of delays in discrete-time domain. Thiran has proposed the time delay approximation algorithm for maximally flat group of fractional delays occurring in signal processing applications. Hence, it is proper candidate for fractional delay compensation for discrete-time SMC design.

The fractional delay in discrete-time can be approximated by Thiran's approximation as under:

$$z^{-\nu} = \Sigma_{k=0}^{l}(-1)^k \binom{l}{k} \Pi_{i=0}^{l} \frac{2\tau'_{sc} + i}{2\tau'_{sc} + k + i} z^{-k}, \tag{4.8}$$

where l indicates the order of approximation and $\nu = \frac{\delta}{h}$ indicates the fractional part of delay, δ is the delay occurring during signal transmission, and h is the sampling interval.

The order of approximation is given by:

$$l = ceil(\nu), \tag{4.9}$$

where *ceil* operator rounds the nearest positive integer greater than or equal to ν. Next, the sliding surface using above approximation is proposed as Lemma 2 given below.

Lemma 2 *The compensated sliding variable $s(k)$ for the given system (4.3, 4.4) with sensor to controller network-induced fractional delay (τ'_{sc}) satisfying condition (4.5) and under the Assumptions (3) and (4) is given as:*

$$s(k) = C_s x(k) - \alpha C_s(x(k-1)), \tag{4.10}$$

where
$\alpha = \frac{\tau'_{sc}}{\tau'_{sc}+1}$ *and C_s is the sliding gain.*

Proof The sliding variable with the delayed state vector at the receiving end of controller is given by:

$$s(k) = C_s x(k - \tau'_{sc}), \tag{4.11}$$

where τ'_{sc} is the sensor to controller fractional delay. The sliding gain C_s is calculated using discrete LQR method through proper selection of Q and R matrices [5].

Applying z-transform to Eq. (4.11), we get

$$s(z) = C_s x(z) z^{-\tau'_{sc}}, \tag{4.12}$$

where $\tau'_{sc} = \frac{\tau_{sc}}{h}$.

$z^{-\tau'_{sc}}$ can be approximated as [4],

$$z^{-\tau'_{sc}} = \Sigma^1_{k=0}(-1)^k \binom{l}{k} \Pi^1_{i=0} \frac{2\tau'_{sc} + i}{2\tau'_{sc} + k + i} z^{-k}. \tag{4.13}$$

The above Eq. (4.13) can be further expanded as,

$$z^{-\tau'_{sc}} = [(-1)^0 \binom{1}{0} \left\{ \frac{2\tau'_{sc}}{2\tau'_{sc}} * \frac{2\tau'_{sc} + 1}{2\tau'_{sc} + 1} \right\} z^0 + (-1)^1 \binom{1}{1} \tag{4.14}$$

$$\left\{ \frac{2\tau'_{sc}}{2\tau'_{sc} + 1} * \frac{2\tau'_{sc} + 1}{2\tau'_{sc} + 2} \right\} z^{-1}].$$

On simplifying, we get

$$z^{-\tau'_{sc}} = 1 - \alpha z^{-1}, \tag{4.15}$$

where $\alpha = \frac{\tau'_{sc}}{\tau'_{sc}+1}$.

Thus, substituting Eq. (4.15) into (4.12),

$$s(z) = C_s x(z)[1 - \alpha z^{-1}]. \tag{4.16}$$

Further expanding, we may get

$$s(z) = C_s x(z) - \alpha C_s z^{-1} x(z). \tag{4.17}$$

Applying inverse z-transform, we may have

$$s(k) = C_s x(k) - \alpha C_s s x(k - 1). \tag{4.18}$$

This completes the *Proof*.

From Eq. (4.18), it is inferred that the network-induced fractional delay from sensor to controller can be compensated in the sliding surface $s(k)$ at each sampling instant h using the current and immediate past sample information and parameter α.

Now we are ready to design a discrete-time sliding mode control law using the proposed sliding surface (4.18).

4.4 Discrete-Time Sliding Mode Control

In this section, non-switching type control law along with its stability is proposed based on reaching law in [6] using compensated sliding surface (4.18). The reaching law proposed by Bartoszewicz causes less chattering compared to Gao's law [7] and offers faster convergence with limited magnitude of the control signal.

Theorem 4.1 *The non-switching discrete-time sliding mode controller for system* (4.3, 4.4) *in the presence of deterministic type sensor to controller fractional delay satisfying condition* (4.5) *and matched uncertainty $d(k)$ is given as,*

$$u(k) = -(C_sG)^{-1}[Hx(k) - Ix(k) - J(s(k)) + d_s(k) - d_1] - d(k). \quad (4.19)$$

where
$H = (C_sF), I = \alpha C_s, J = \{1 - q[s(k)]\}.$

Proof Let us consider the reaching law in [6] in the presence of sensor to controller fractional delay given as:

$$s[(k + 1)h] = \{1 - q[s(k)]\} - d_s(k) + d_1, \quad (4.20)$$

where $d_s(k)$ is disturbance at the controller end, $q[s(k)] = \frac{\psi}{\psi + |s(k)|}$ with ψ as user defined constant satisfying $\psi \geq d_2$, d_1 and d_2 are mean and deviation of $d(k)$.

Remark 8 The disturbance $d(k)$ appearing in the reaching law is applied through the network. So, the compensated disturbance using Thiran's approximation is given as:

$$d_s(k) = d(k) - \alpha d(k - 1). \quad (4.21)$$

The reaching law in Eq. (4.20) indicates that the system states always move towards the specified sliding band given as:

$$|s(k)| \leq \frac{\psi d_2}{\psi - d_2}. \quad (4.22)$$

Substituting Eq. (4.18) into Eq. (4.20), we get

$$C_sx(k + 1) - \alpha C_sx(k) = \{1 - q[s(k)]\} - d_s(k) + d_1,$$

Remark 9 It is noticed from Eq. (4.18) that the sensor to controller fractional delay is compensated at the sliding surface, while controller to actuator fractional delay is compensated at the actuator side. So, the effect of delay will not be observed at the controller side for computing the control signal as well at actuator side. Thus, without loss of generality, the control signal in Eq. (4.3) is given as

$$u(k - \tau') = u(k) \quad (4.23)$$

Further, substituting $x(k + 1)$, we get

$$C_s[Fx(k) + Gu(k) + d(k)] - \alpha C_sx(k) = \{1 - q[s(k)]\} - d_s(k) + d_1.$$

Further simplification gives,

$$C_s Fx(k) + C_s Gu(k) + C_s d(k) - \alpha C_s x(k) = \{1 - q[s(k)]\} - d_s(k) + d_1. \quad (4.24)$$

Solving the above Eq. (4.24), control law is expressed as

$$u(k) = -(C_s G)^{-1}[Hx(k) - Ix(k) - J(s(k)) + d_s(k) - d_1] - d(k). \quad (4.25)$$

where

$$H = (C_s F), I = \alpha C_s \text{ and } J = \{1 - q[s(k)]\}$$

This completes the *proof*.

Next, the stability condition is derived using compensated sliding surface (4.18) and control law proposed in Eq. (4.25) such that the system states remain within specified band (4.22) over a finite interval of time.

4.5 Stability Analysis

Theorem 4.2 *For given positive scalars τ'_{sc} and τ'_{ca} with total networked delay (τ'), the trajectories of the closed-loop system (4.3, 4.4) with controller (4.25) and $d(k)$ satisfying (4.7) drive towards the sliding surface (4.18) provided following condition (4.26) is feasible:*

$$\eta s^T(k)s(k) \succ 0. \quad (4.26)$$

Proof The compensated sliding surface is given by,

$$s(k) = C_s x(k) - \alpha C_s x(k-1). \quad (4.27)$$

Selecting the Lyapunov function as,

$$V_s(k) = s^T(k)s(k). \quad (4.28)$$

Writing forward difference of the above equation,

$$\Delta V_s(k) = s^T(k+1)s(k+1) - s^T(k)s(k). \quad (4.29)$$

Substituting the value of $s(k+1)$ using Eq. (4.27), we get

$$\Delta V_s(k) = [C_s x(k+1) - \alpha C_s x(k)]^T [2C_s x(k+1) \\ -\alpha C_s x(k)] - s^T(k)s(k). \quad (4.30)$$

Substituting the value of $x(k + 1)$,

$$\Delta V_s(k) = [C_s[Fx(k) + G(u(k) + d(k))] - \alpha C_s x(k)]^T \tag{4.31}$$
$$[C_s Fx(k) + G(u(k) + d(k))] - \alpha C_s x(k)] - s^T(k)s(k).$$

Substituting the value of $u(k)$ from Eq. (4.25) and further solving it, we have

$$\Delta V_s(k) = [(1 - q[s(k)])s(k) - d_s(k) + d_1]^T * [(1 - q[s(k)]) \tag{4.32}$$
$$s(k) - d_s(k) + d_1] - s^T(k)s(k).$$

Denoting,

$$\kappa = [(1 - q[s(k)])s(k) - d_s(k) + d_1]^T * [(1 - q[s(k)])s(k) - d_s(k) + d_1]$$

Then, we have

$$\Delta V_s(k) = \kappa - s^T(k)s(k). \tag{4.33}$$

The term κ can be tuned close to zero by appropriately selecting the parameter ψ. If κ is closed to zero, then $s^T(k)s(k)$ will be larger than κ. Thus, for any small parameter η, we have $\kappa - s^T(k)s(k) \prec \eta s^T(k)s(k)$.

So, by tuning the parameter ψ, we have $\Delta V_s(k) \prec \eta s^T(k)s(k)$ which guarantees the convergence of $\Delta V_s(k)$ and implies that any trajectory of the system (4.3, 4.4) will be driven onto the sliding surface and maintain on it.
This completes the *proof*.

4.6 Results and Discussions

This section briefly discusses about the simulation results as well as experimental results of the proposed control algorithm in the presence of deterministic network delays and matched uncertainty. The efficiency and robustness of the proposed control algorithms are tested under three different situations: (i) illustrative example (ii) real-time plant as DC servo motor and (iii) real-time networks.

4.6.1 Illustrative Example

In this section, an illustrative example given in [8] is simulated in MATLAB environment.

Consider the continuous-time LTI system as,

$$\dot{x}(t) = Ax(t) + Bu(t - \tau) + Dd(t), \tag{4.34}$$
$$y(t) = Cx(t), \tag{4.35}$$

where
$$A = \begin{bmatrix} -0.7 & 2 \\ 0 & -1.5 \end{bmatrix}, B = \begin{bmatrix} -0.03 \\ -1 \end{bmatrix},$$
$$C = \begin{bmatrix} 1 & 0 \end{bmatrix}, D = \begin{bmatrix} 1 \\ 1 \end{bmatrix}, d(t) = 0.2\sin(0.086t).$$

Discretizing the above system parameters at sampling interval of $h = 30\,\text{ms}$,

$$x(k+1) = Fx(k) + Gu(k - \tau') + d(k), \tag{4.36}$$
$$y(k) = Cx(k), \tag{4.37}$$

where
$$F = \begin{bmatrix} 0.9792 & 0.05805 \\ 0 & 0.956 \end{bmatrix}, G = \begin{bmatrix} -0.001771 \\ -0.02934 \end{bmatrix},$$
$$C = \begin{bmatrix} 1 & 0 \end{bmatrix}.$$

Figures 4.3, 4.4, 4.5, 4.6, 4.7, 4.8, 4.9, 4.10, 4.11, 4.12, 4.13 and 4.14 shows the nature of the system under networked environment. In order to check the robustness of the derived control law, a slowly time-varying disturbance is applied at the input of the system as shown in Fig. 4.2. The deterministic type network-induced fractional delay with range of $3\,\text{ms} \le \tau \le 20\,\text{ms}$ is shown in Fig. 4.3. In this work, network delay is considered as the time required for the data packets to travel from sensor to controller and controller to actuator. The time required for data packets to travel from sensor to controller is $1.5\,\text{ms} \le \tau_{sc} \le 10\,\text{ms}$ and for controller to actuator is $1.5\,\text{ms} \le \tau_{ca} \le 10\,\text{ms}$, respectively. The sliding gain C_s is calculated using discrete LQR method with $Q = diag(1000, 1000)$ and $R = 1$. The computed values of sliding gain are $C_s = [-1.77 \ -2.766]$. The quasi-sliding mode band computed to be $|s(k)| \le +0.2$ to -0.2 with proper selection of user-defined constant $\psi = 100$.

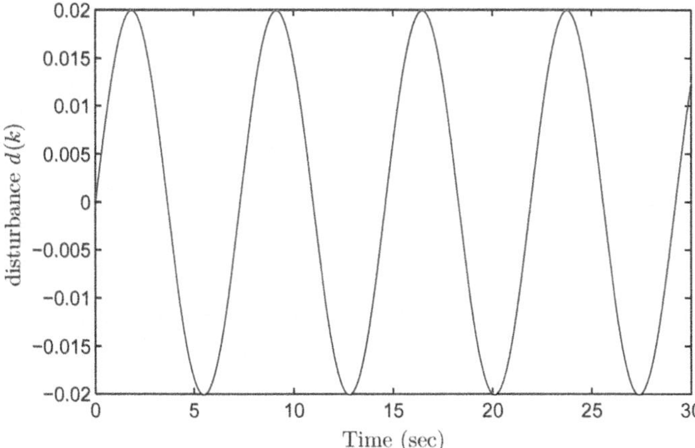

Fig. 4.2 Slowly time-varying disturbance $d(k)$

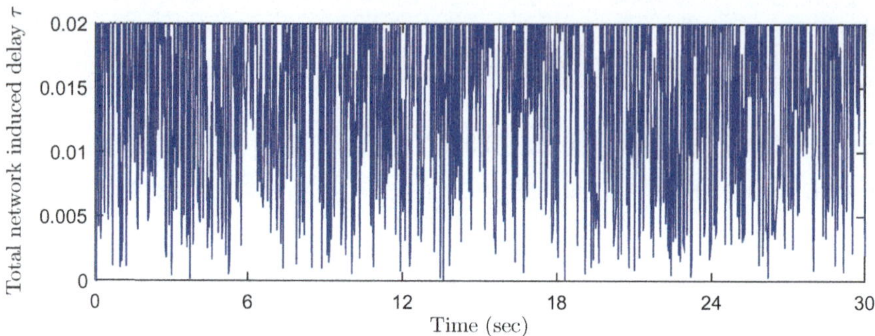

Fig. 4.3 Total network delay τ

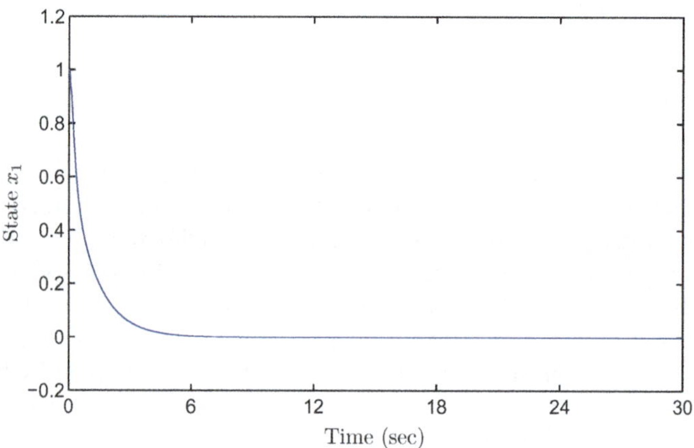

Fig. 4.4 State variable x_1 with initial condition $x_1 = 1$

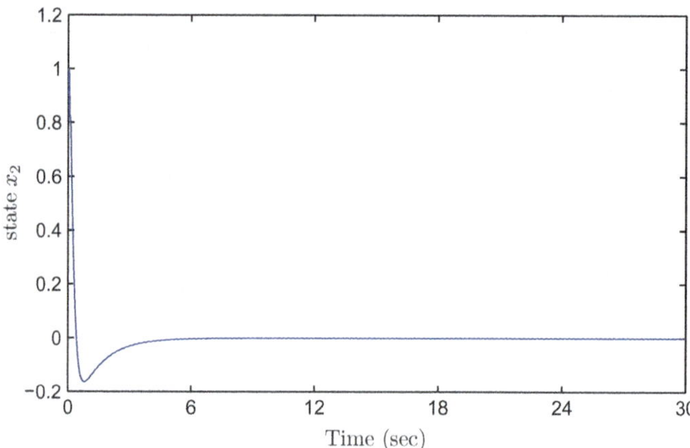

Fig. 4.5 State variable x_2 with initial condition $x_2 = 1$

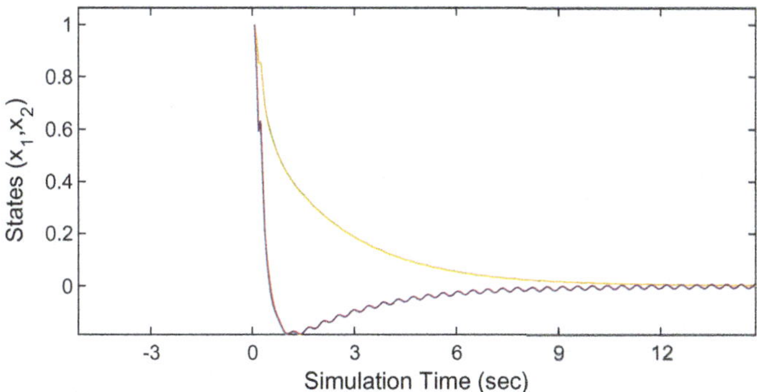

Fig. 4.6 Magnified result of state variables x_1, x_2

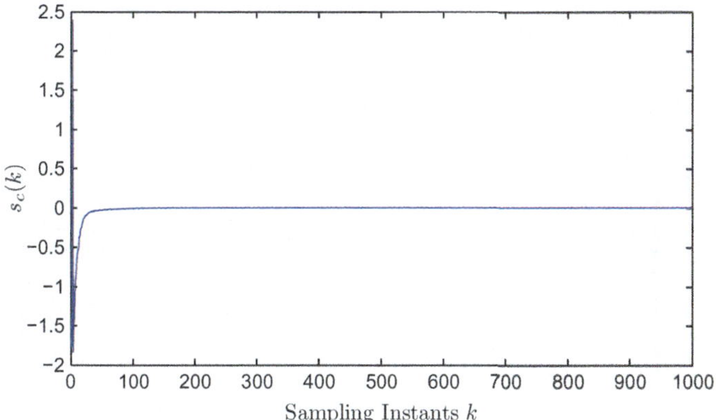

Fig. 4.7 Compensated sliding surface $s(k)$

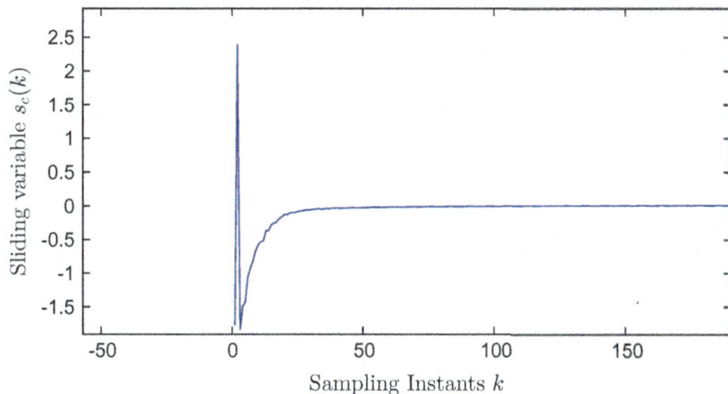

Fig. 4.8 Magnified result of compensated sliding surface $s(k)$

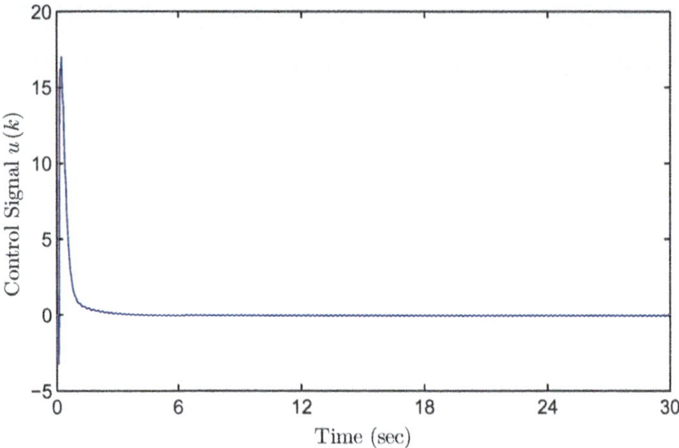

Fig. 4.9 Control signal $u(k)$

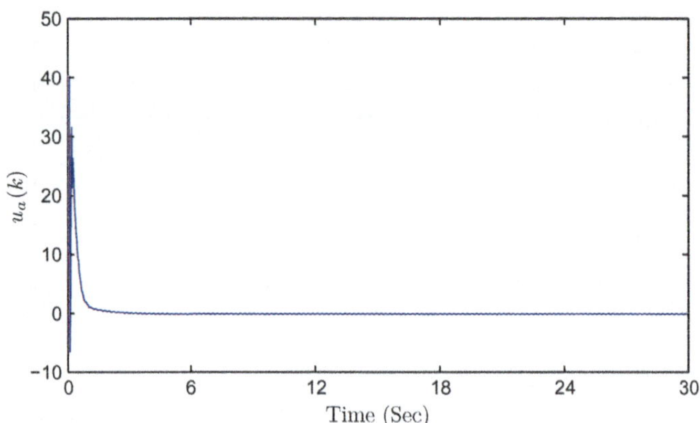

Fig. 4.10 Compensated control signal $u_a(k)$

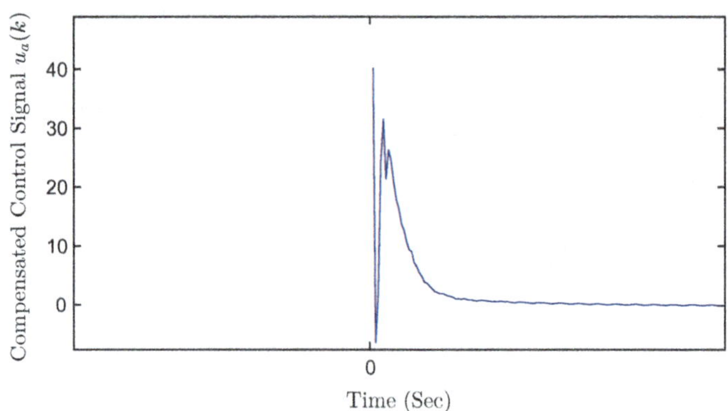

Fig. 4.11 Magnified compensated control signal $u_a(k)$

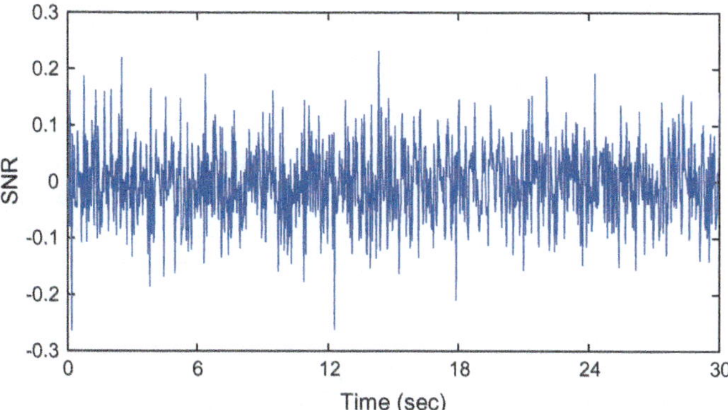

Fig. 4.12 Response of SNR

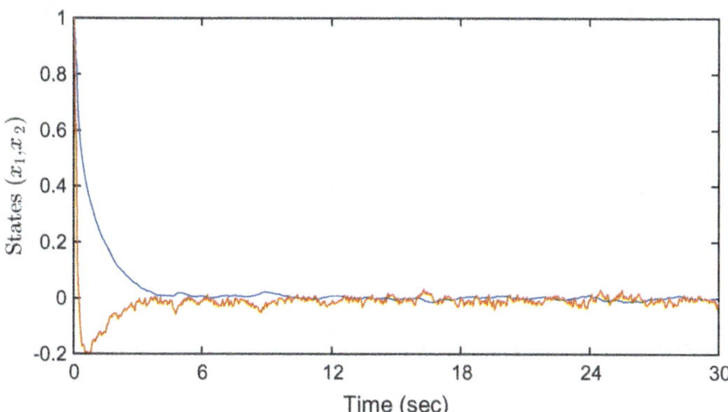

Fig. 4.13 Nature of state variables for different SNR

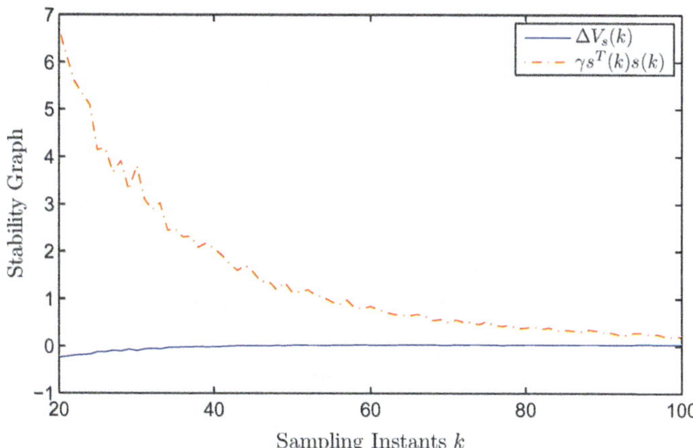

Fig. 4.14 Result of stability

Figures 4.4 and 4.5 show the plant state variables with initial condition $x(k) = [1\ 1]$. Both the states converge to zero from given initial condition in the presence of network fractional delay. Figure 4.6 shows the magnified response of the plant state variables. It can be noticed that the effect of fractional delay is compensated from first sampling instant. Figure 4.7 shows the compensated sliding surface calculated using Thiran's approximation. It can be observed that the compensated sliding variable is also computed from first sampling instant in the presence of sensor to controller fractional delay. The magnified response of the same is shown in Fig. 4.8. Figure 4.9 shows the control signal $u(k)$ which is computed using proposed compensated sliding surface $s(k)$. This control signal is further applied to the plant through the network. The same approach of time delay compensation is used to compute the compensated control signal $u_a(k)$ as shown in Fig. 4.10. From the magnified result in Fig. 4.11, it can be justified that the effect of controller to actuator fractional delay is also compensated from first sampling interval.

The algorithm is also examined for different signal-to-noise ratio (SNR) as shown in Fig. 4.12. It can be observed from Fig. 4.13 that the system states converge to zero for different SNR. Figure 4.14 shows the results of stability. It is observed from Fig. 4.14 that for given $\psi = 100$ and $d_2 = 0.2$ guarantees the covergence of $\Delta V_s(k)$ and implies that the trajectories of system (3.3, 3.4) will be driven on the compensated sliding surface and maintain on it under the specified network fractional delay and matched uncertainty.

4.6.2 Simulation and Experimental Results of Brushless DC Motor

The state space model of the system (3.35) is given as,

$$\dot{x}(t) = Ax(t) + Bu(t - \tau) + Dd(t), \qquad (4.38)$$

$$y(t) = Cx(t), \qquad (4.39)$$

where

$$A = \begin{bmatrix} -201 & 0 \\ 1 & 0 \end{bmatrix}, B = \begin{bmatrix} 1 \\ 0 \end{bmatrix},$$

$$C = \begin{bmatrix} 0 & 1 \end{bmatrix}, D = \begin{bmatrix} 1 \\ 1 \end{bmatrix}, d(t) = 0.2\sin(0.086t).$$

Discretizing the system at sampling interval $h = 30$ ms, we get

$$x(k + 1) = Fx(k) + Gu(k - \tau') + d(k), \qquad (4.40)$$

$$y(k) = Cx(k), \qquad (4.41)$$

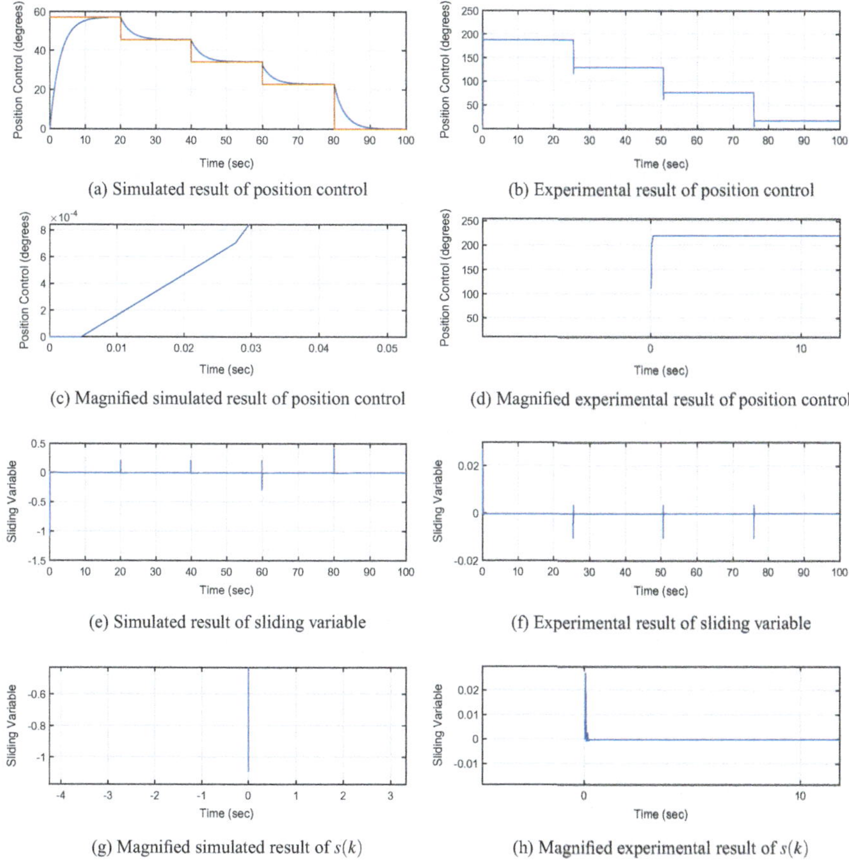

(a) Simulated result of position control

(b) Experimental result of position control

(c) Magnified simulated result of position control

(d) Magnified experimental result of position control

(e) Simulated result of sliding variable

(f) Experimental result of sliding variable

(g) Magnified simulated result of $s(k)$

(h) Magnified experimental result of $s(k)$

Fig. 4.15 Simulation and experimental result of position tracking and sliding variable for $\tau = 10\,\text{ms}$

where
$$F = \begin{bmatrix} 0.001836 & 0 \\ 0.004753 & 1 \end{bmatrix}, G = \begin{bmatrix} 0.004753 \\ -0.0001242 \end{bmatrix},$$

$$C = \begin{bmatrix} 0 & 1 \end{bmatrix}.$$

This section briefs about the simulation and experimental results of position control brushless DC motor in the presence of various deterministic delays using non-switching control law. The effect of time delay compensation is deeply analysed through tracking response, compensated sliding variable and control signal for different network delays as shown in Figs. 4.14, 4.15, 4.16, 4.17 and 4.18. The robustness of the proposed algorithm is determined by applying time-varying disturbance signal at the input side of the channel. The total networked-induced delay with a range of 10–28 ms was generated for which the effect of time delay is compensated satisfying condition (3.15). The sliding gain is computed through discrete LQR method which comes out to be $C_s = [24.5156\ 31.6288]$ with $Q = diag(1000, 1000)$ and

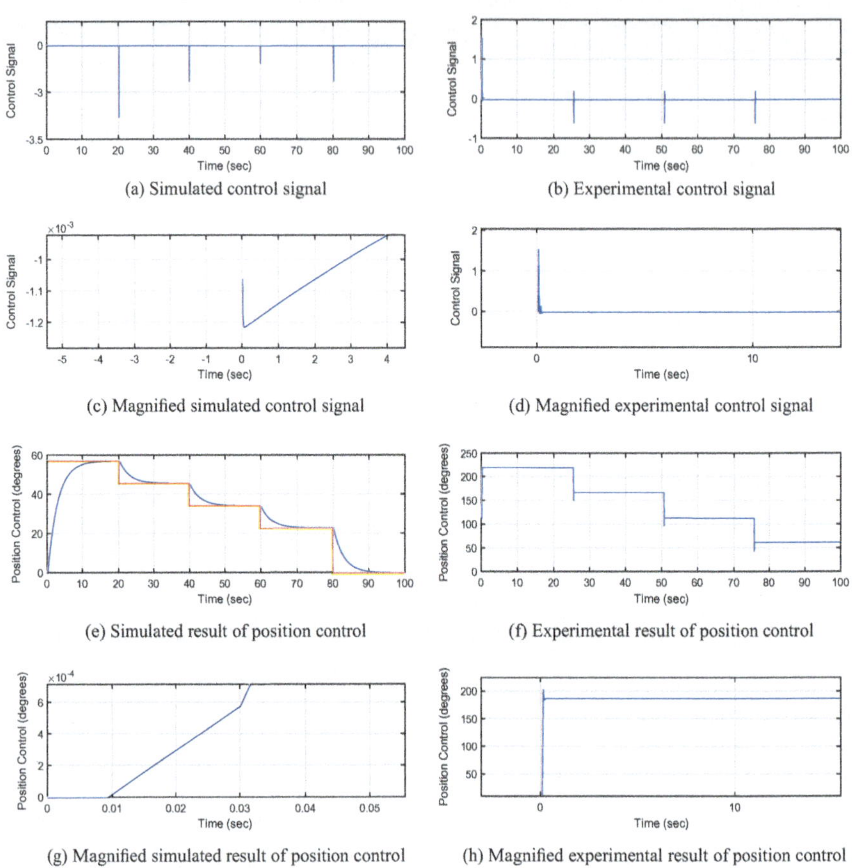

(a) Simulated control signal

(b) Experimental control signal

(c) Magnified simulated control signal

(d) Magnified experimental control signal

(e) Simulated result of position control

(f) Experimental result of position control

(g) Magnified simulated result of position control

(h) Magnified experimental result of position control

Fig. 4.16   **a–d** Simulation and experimental result of control signal for $\tau = 18$ ms and **e–h** simulation and experimental result of position tracking for $\tau = 24$ ms

$R = 1$. The quasi-sliding mode band computed to be $|s(k)| \leq +0.2$ to -0.2 with proper selection of user-defined constant $\psi = 1500$. Figures 4.14a to 4.15d show the simulated and experimental results of position control brushless DC motor for total networked-induced delay of $\tau = 10$ ms with $\tau_{sc} = 5$ ms and $\tau_{ca} = 5$ ms. The fractional part of total networked delay for sampling interval of $h = 30$ ms is computed to be $\tau' = 0.33$, $\tau'_{sc} = 0.166$ and $\tau'_{ca} = 0.166$, respectively. Figure 4.14a, b shows the simulated as well as experimental trajectory results of the plant. It can be observed that the position of DC motor is controlled according to variations in the reference inputs without chattering even in the presence of specified network delay. The tracking results are magnified as shown in Fig. 4.14c, d in order to examine the effect of time delay compensation. It can be noticed that in both the cases, the fractional part of the delay from sensor to controller is compensated as the position of the motor commences the reference input signal at 5 ms. The same conse-

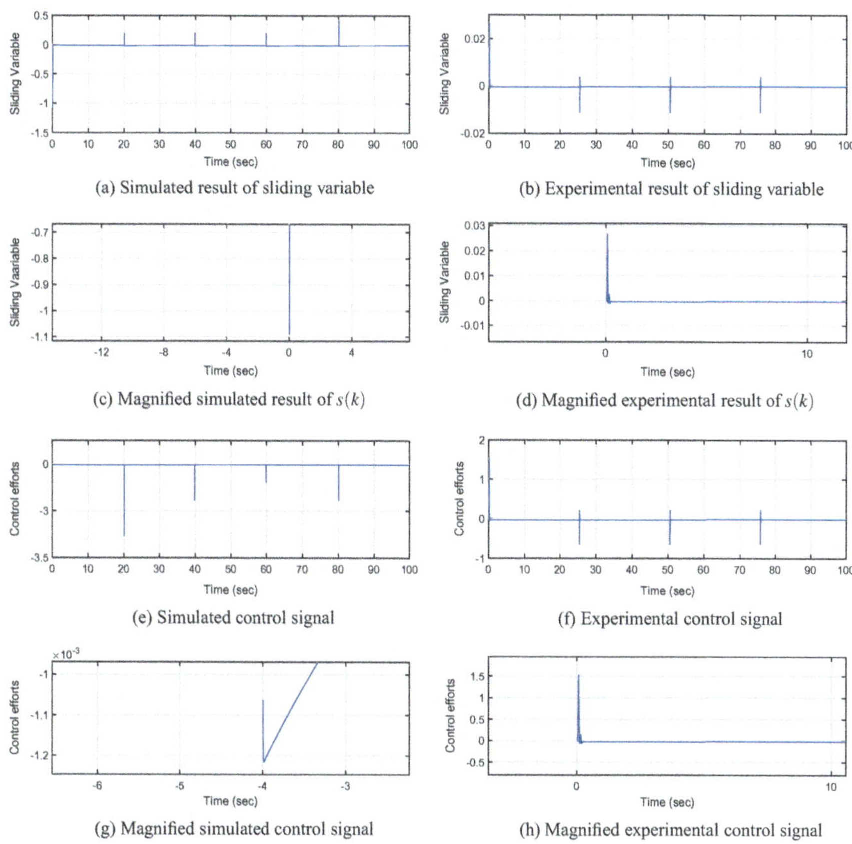

(a) Simulated result of sliding variable

(b) Experimental result of sliding variable

(c) Magnified simulated result of $s(k)$

(d) Magnified experimental result of $s(k)$

(e) Simulated control signal

(f) Experimental control signal

(g) Magnified simulated control signal

(h) Magnified experimental control signal

Fig. 4.17 Simulation and experimental result of sliding variable and control signal for $\tau = 18\,\text{ms}$

quence of time delay compensation is observed in sliding variable, Fig. 4.14e, f, as well as control signal, Fig. 4.15a, b, respectively. Observing the magnified results, Figs. 4.14g, h and 4.15c, d, it can be noticed that the compensated sliding variable and control signal both are computed from first sampling instants. Thus, the effect of fractional delay from sensor to controller is compensated at the sliding surface and control signal. The proposed algorithm was further examined for $\tau = 18\,\text{ms}$ and $28\,\text{ms}$, respectively. Figures 4.15e–h shows the results of position control of brushless DC motor. The results are carried out under the total networked delay of $\tau = 18\,\text{ms}$ with $\tau_{sc} = 9\,\text{ms}$ and $\tau_{ca} = 9\,\text{ms}$. The fractional part of delay for $h = 30\,\text{ms}$ is computed as $\tau' = 0.6$, $\tau'_{sc} = 0.3$ and $\tau'_{ca} = 0.3$, respectively. The simulated and experimental tracking results of the plant for the specified networked delay are shown in Fig. 4.15e, f. It can be noticed that in both the cases, the position of DC motor is controlled for all given reference inputs. In order to examine the effect of fractional time delay compensation, the same results are magnified in Fig. 4.15g, h. It can be noticed that the effect of fractional delay from sensor to controller is nullified as

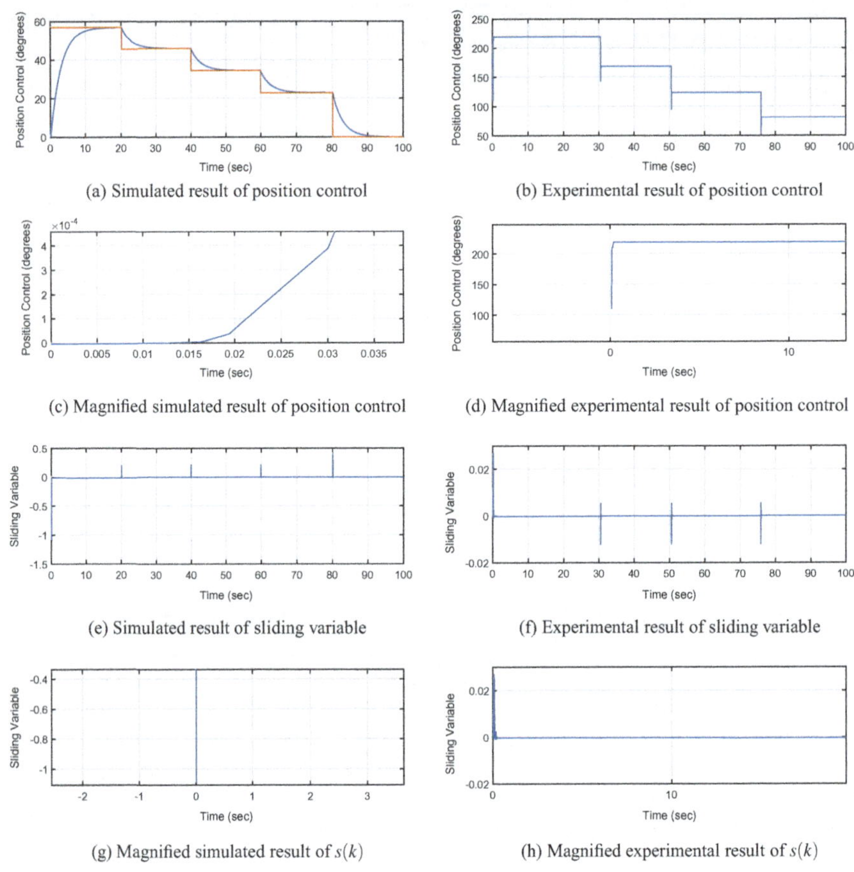

(a) Simulated result of position control

(b) Experimental result of position control

(c) Magnified simulated result of position control

(d) Magnified experimental result of position control

(e) Simulated result of sliding variable

(f) Experimental result of sliding variable

(g) Magnified simulated result of $s(k)$

(h) Magnified experimental result of $s(k)$

Fig. 4.18   Simulation and experimental result of position tracking and sliding surface for $\tau = 28$ ms

the output follows the reference signal at $t = 9$ ms. The same effect of time delay compensation is observed in sliding variable, Fig. 4.16a, b, as well as control signal, Fig. 4.16e, f. Observing the magnified results, Fig. 4.16c, d, g, h, it can be noticed that in both the cases, the sliding variable and control signal are computed from first sampling instants. Thus, the effect of fractional part of delay from sensor to controller is compensated at sliding variable and control signal. Figures 4.17a and 4.18d shows the results of position control of brushless DC motor for total networked delay of $\tau = 28$ ms with $\tau_{sc} = 14$ ms and $\tau_{ca} = 14$ ms. Considering the sampling interval of $h = 30$ ms, the fractional part of delay is computed as $\tau' = 0.933$, $\tau'_{sc} = 0.466$ and $\tau'_{ca} = 0.46$ respectively. Figure 4.17a, b show the simulated and experimental tracking results of the system for the specified networked delay. It can be observed that in both the cases, output tracks the reference trajectory without chattering. In order to examine the actual effect of time delay compensation, the results are magnified as shown in Fig. 4.17c, d. Observing the magnified results, it can be concluded that the effect of fractional part of delay from sensor to controller is compensated as output tracks the reference signal at $t = 14$ ms. Figures 4.17e, f and 4.18a, b show the simulated and experimental results of compensated sliding variable and control signal for specified networked delay. It can be noticed from the results that the time delay compensation algorithm works efficiently for large value of τ. The magnified results of the same are shown in Figs. 4.17g, h and 4.18c, d which clearly justifies that in both the cases, the effect of fractional part of delay from sensor to controller is compensated in the sliding variable as well as control signal.

Thus, from above results, it can be concluded that the non-switching controller designed using proposed algorithm works efficiently with the network delay range of 10–28 ms in simulated as well as experimental environment. The proposed controller compensates the effect of fractional time delay without chattering for $\psi = 1500$ satisfying (4.22) and shows the stable response satisfying (4.26) in the presence of matched uncertainty.

4.6.2.1 Comparison of Proposed Algorithm with Conventional Sliding Mode Control

In this section, the experimental results of proposed algorithm are compared with switching sliding mode control using time delay approximation and conventional sliding mode control. The results are compared in terms of tracking response, sliding variable and control signal for total networked delay of $\tau = 10$ ms. From the comparative results (Figs. 4.19a and 4.20), it can be observed that the conventional sliding mode control becomes unstable for a small delay of $\tau_{sc} = 5$ ms while the sliding mode controller designed using switching algorithm generates the chattering behaviour at the output signal compared to proposed algorithm. Thus, Thiran approximation proved to be more efficient technique for non-switching-based discrete-time

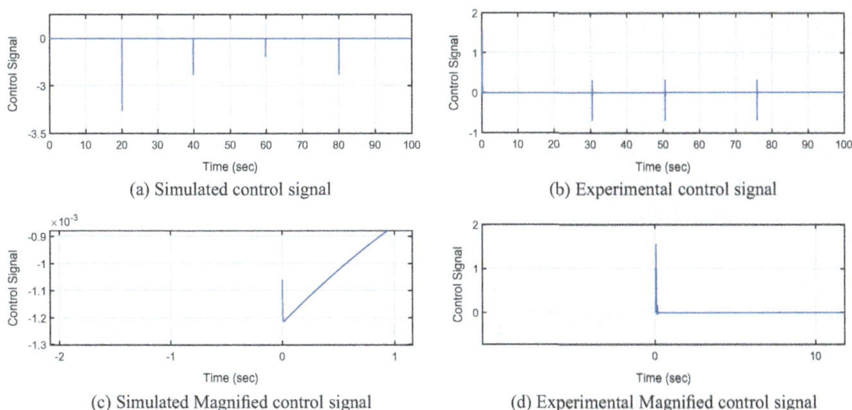

(a) Simulated control signal

(b) Experimental control signal

(c) Simulated Magnified control signal

(d) Experimental Magnified control signal

Fig. 4.19 Simulation and experimental result of control signal for $\tau = 28$ ms

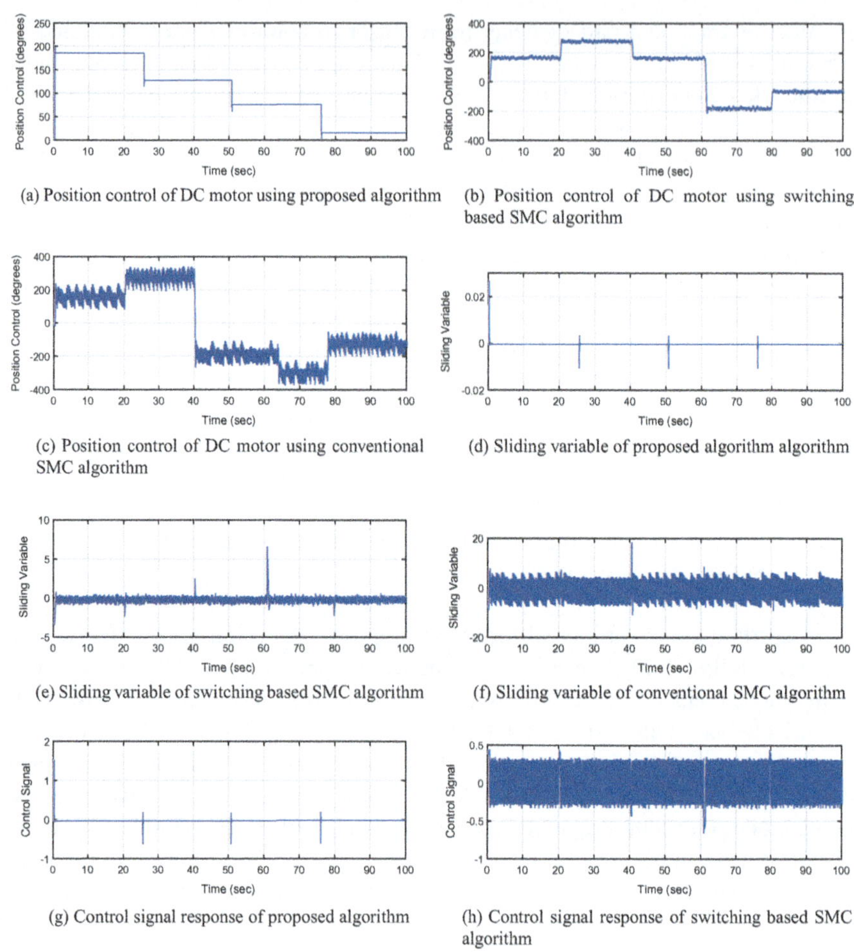

(a) Position control of DC motor using proposed algorithm

(b) Position control of DC motor using switching based SMC algorithm

(c) Position control of DC motor using conventional SMC algorithm

(d) Sliding variable of proposed algorithm algorithm

(e) Sliding variable of switching based SMC algorithm

(f) Sliding variable of conventional SMC algorithm

(g) Control signal response of proposed algorithm

(h) Control signal response of switching based SMC algorithm

Fig. 4.20 Comparison of proposed algorithm, switching based sliding mode control and conventional sliding mode control for $\tau = 12.8$ ms

sliding mode control. The comparison of proposed algorithm with switching sliding mode control using time delay approximation and conventional sliding mode control is summarized in Table 4.1 (Fig. 4.21).

4.7   Simulation with Real-Time Networks

In previous section, the efficacy of the proposed control law is examined in the presence of brushless DC motor connected through networked medium. It can be observed from simulation results that the control law proposed using non-switching

Table 4.1 Comparison of proposed algorithm, switching-based SMC and conventional SMC

Algorithm	Comparative results			
	τ (ms)	T_s	Chattering	Response
Conventional SMC	12.8	Undefined	High	Unstable
Switching SMC	12.8	1 s	Within QSMB	Stable
Proposed method	12.8	0.3 s	Nil	Stable

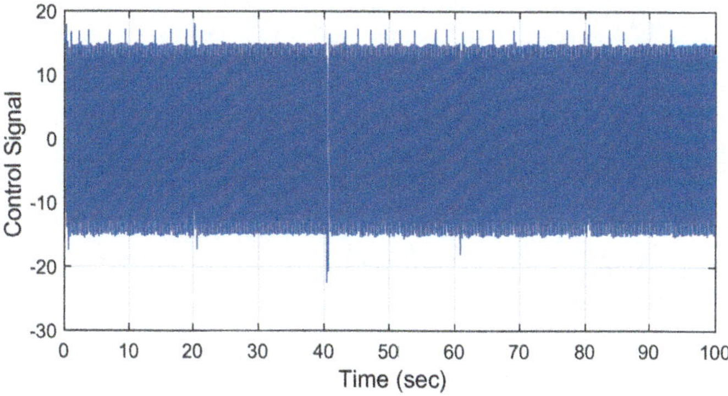

Fig. 4.21 Control signal response of conventional SMC algorithm for $\tau = 12.8$ ms

reaching law provides faster convergence without increasing the amplitude of control signal. The chattering is also negligible compared to switching type control law. Thus, in this section, the efficacy of the proposed non-switching control law is further tested in the presence of real-time networks and matched uncertainty. The real-time networks are simulated using true time software which provides wide range of simulated networks such as CAN, Switched Ethernet, Profibus, Profinet, CSMA/CD, Round Robbin [9]. In this work, the simulations are carried out under CAN and Switched Ethernet communication medium as network delays are assumed to be deterministic in nature. Further, the performance of the system is also studied in the presence of packet loss situation.

The following network specifications and parameters are considered for simulations:

Networked medium: CAN and Switched Ethernet
Data rate (bits/s) = 80,000
Minimum frame size (bits) = 512 (*CAN*) and 1024 (*Switched Ethernet*)
Loss probability = 0 to 0.5
sampling interval h = 0.030 s.
$C_s = [24.5156\ 31.6288]$
$|s(k)| \leq +0.2$ to -0.2 with user-defined constant $\psi = 1500$.

Consider the continuous-time LTI system as,

$$\dot{x}(t) = Ax(t) + Bu(t - \tau) + Dd(t), \tag{4.42}$$

$$y(t) = Cx(t), \tag{4.43}$$

where
$$A = \begin{bmatrix} -0.7 & 2 \\ 0 & -1.5 \end{bmatrix}, B = \begin{bmatrix} -0.03 \\ -1 \end{bmatrix},$$

$$C = \begin{bmatrix} 1 & 0 \end{bmatrix}, D = \begin{bmatrix} 1 \\ 1 \end{bmatrix}, d(t) = 0.2\sin(0.086t).$$

Discretizing the above system parameters at sampling interval of $h = 30\,\text{ms}$, we get

$$x(k + 1) = Fx(k) + Gu(k - \tau') + d(k), \tag{4.44}$$

$$y(k) = Cx(k), \tag{4.45}$$

where
$$F = \begin{bmatrix} 0.9792 & 0.05805 \\ 0 & 0.956 \end{bmatrix}, G = \begin{bmatrix} -0.001771 \\ -0.02934 \end{bmatrix},$$

$$C = \begin{bmatrix} 1 & 1 \end{bmatrix}.$$

4.7.1 CAN as a Network Medium

In this section, the nature of the system with CAN as a networked medium is studied in
Figs. 4.22, 4.23, 4.24, 4.25, 4.26, 4.27, 4.28, 4.29, 4.30, 4.31, 4.32, 4.33, 4.34, 4.35,
4.36 and 4.37. The robustness of the proposed controller is checked by applying
slowly time-varying disturbance at the input side of the system as shown in Fig. 4.2.
In CAN, it is assumed that the minimum frame size is 512 bits and data transfer
rate is 80,000 bits/s. So, the delay generated in the CAN network to transfer the data
packets from sensor to controller is $\tau_{sc} = 6.4\,\text{ms}$ and from controller to actuator is

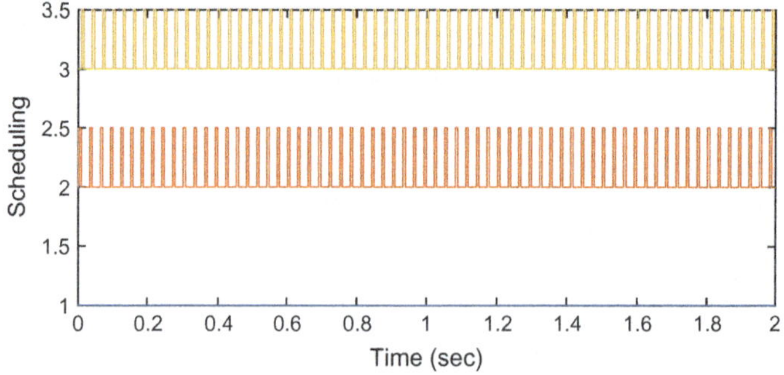

Fig. 4.22 Scheduling policies of sensor to actuator with CAN network under ideal condition

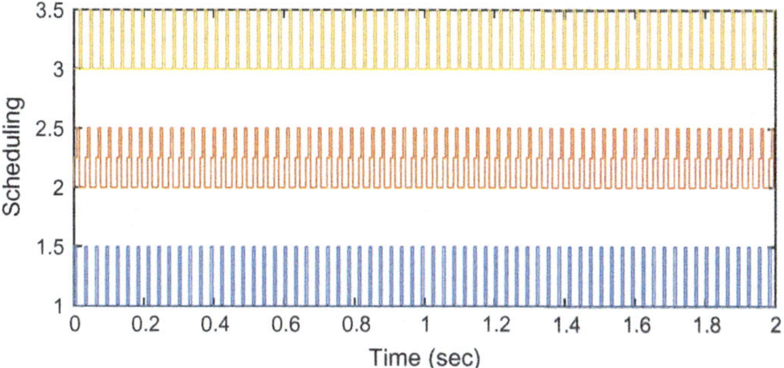

Fig. 4.23 Scheduling policies of sensor to actuator with CAN network under traffic conditions

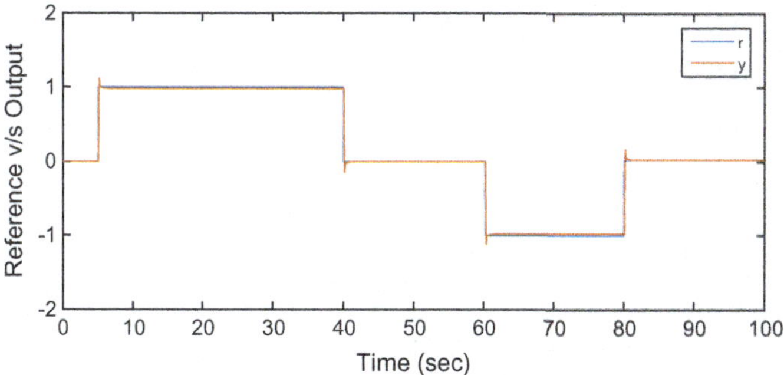

Fig. 4.24 Tracking response of system with CAN network under ideal condition

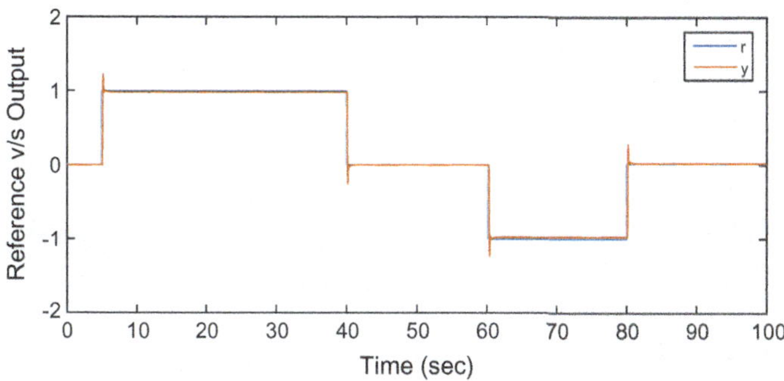

Fig. 4.25 Tracking response of system with CAN network under traffic conditions

$\tau_{ca} = 6.4$ ms. The processing and the computational delays at sensor, controller and actuator are considered as 0.9, 0.5 and 0.5 ms, respectively. Thus, the total networked delay generated within the closed-loop system is $\tau = 14.7$ ms. The fractional part of total network delay is obtained as $\tau' = 0.49$, $\tau'_{sc} = 0.213$ and $\tau'_{ca} = 0.213$ for sampling interval of $h = 30$ ms. The scheduling policies of sensor to controller and

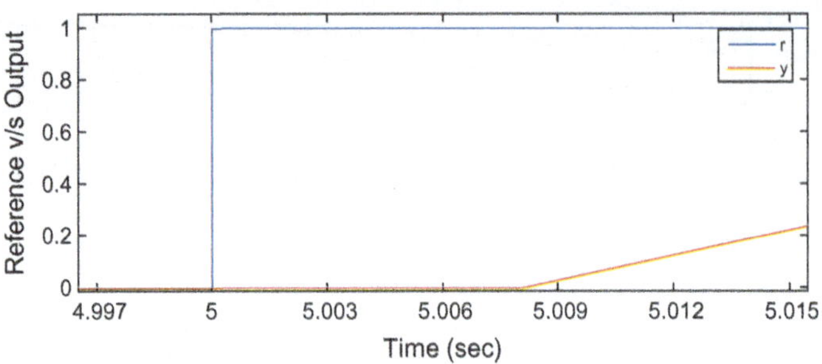

Fig. 4.26 Magnified tracking response with CAN network under ideal conditions

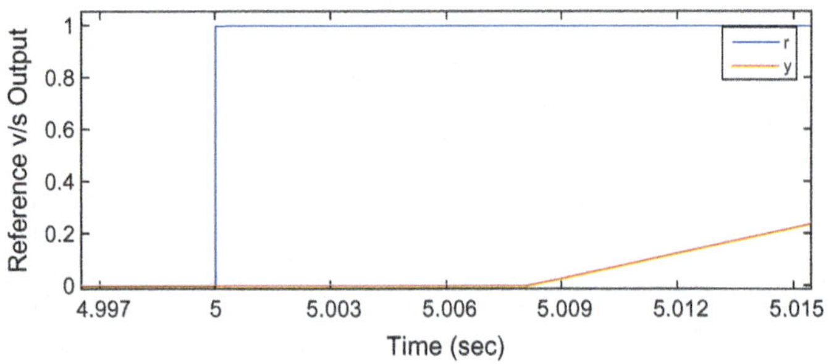

Fig. 4.27 Magnified tracking response with CAN network under traffic conditions

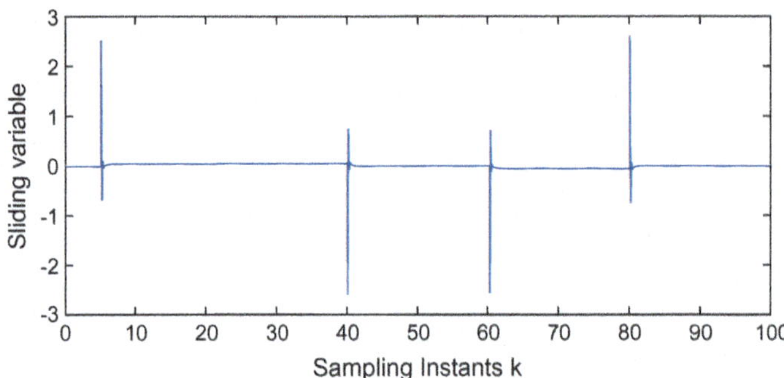

Fig. 4.28 Compensated sliding variable $s(k)$

controller to actuator with network under ideal condition and bandwidth sharing condition are shown in Figs. 4.22 and 4.23, respectively. It can be observed that blue samples are indicated as the traffic while yellow and red samples indicate the scheduling policy for sensor to controller and controller to actuator. The trajectory response of the system for the network under ideal condition and traffic condition is shown in Figs. 4.24 and 4.25. It can be noticed that under both situations, the output tracks the reference trajectory for the specified networked delay. In order to show the precise effect of time delay compensation in CAN network at the output results

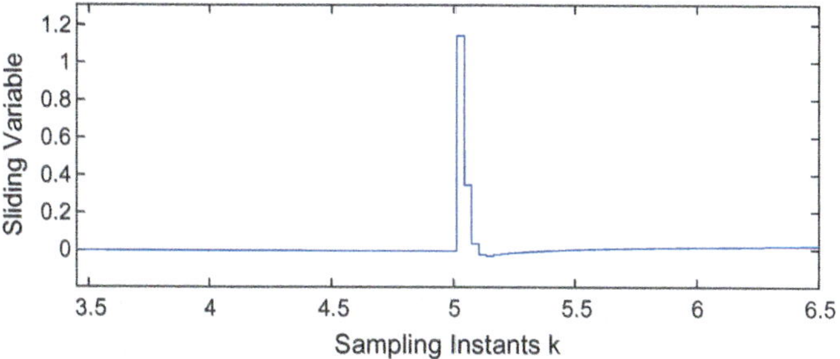

Fig. 4.29 Magnified compensated sliding variable $s(k)$

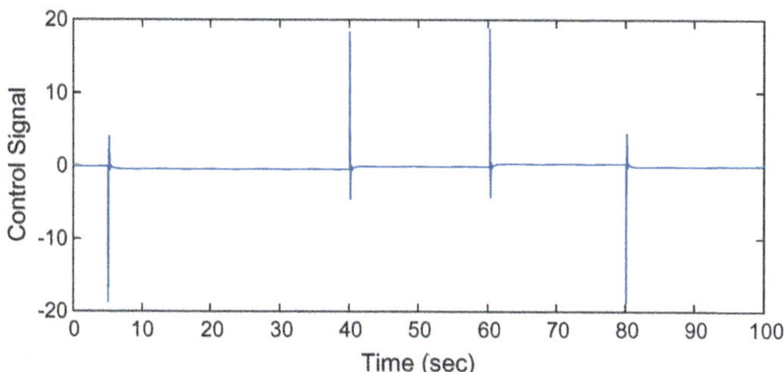

Fig. 4.30 Control signal $u(k)$

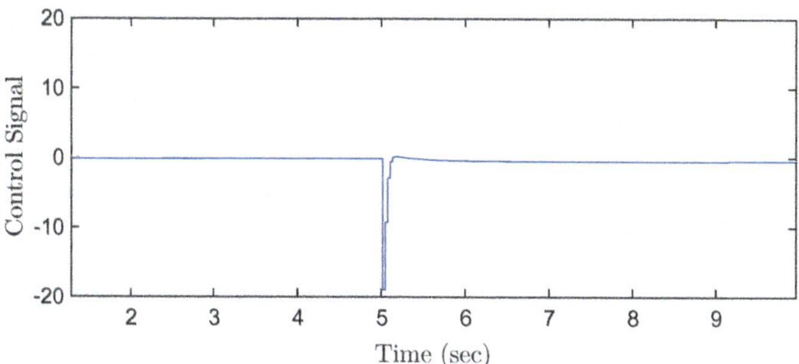

Fig. 4.31 Magnified control signal $u(k)$

are magnified as shown in Figs. 4.26 and 4.27. It can be noticed that the effect of fractional delay from sensor to controller is compensated as the output tracks the reference input at $t = 8.3$ ms. The similar effect of time delay compensation for the network under traffic condition can be observed in sliding variable, Figs. 4.28 and 4.29, as well as control signal, Figs. 4.30 and 4.31. Observing the magnified results in Figs. 4.29 and 4.31, it can be noticed that the sliding variable and control signal are computed exactly after an interval of $t = 1.4$ ms even in the presence of sensor to

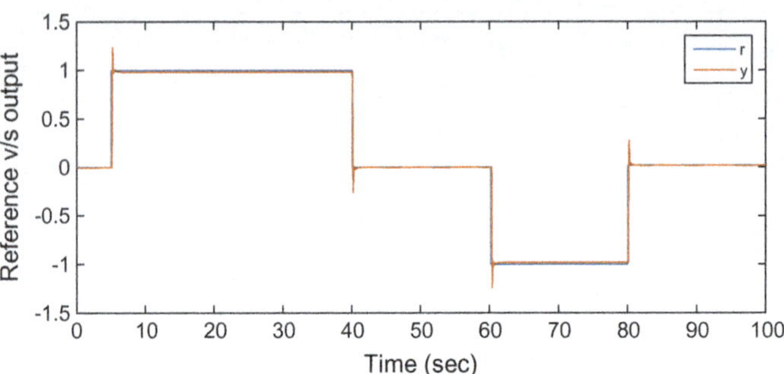

Fig. 4.32 Tracking response of system when packet loss is 10%

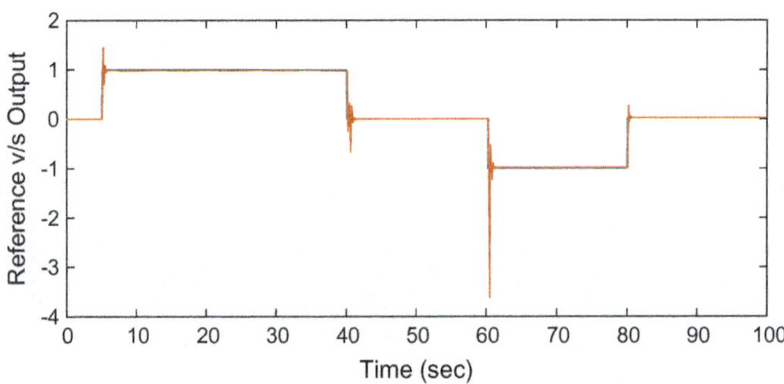

Fig. 4.33 Tracking response of system when packet loss is 30%

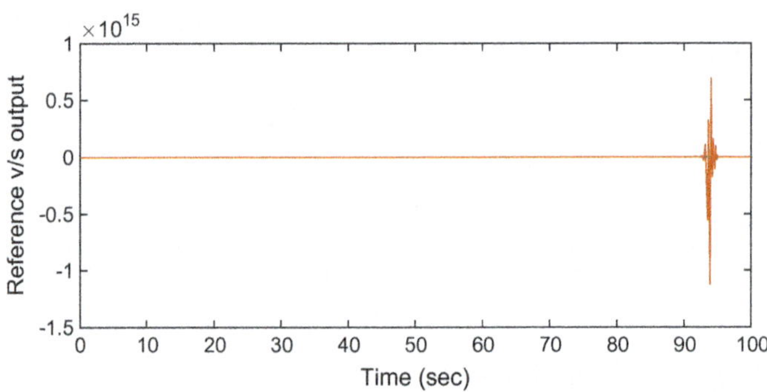

Fig. 4.34 Tracking response of system when packet loss is 50%

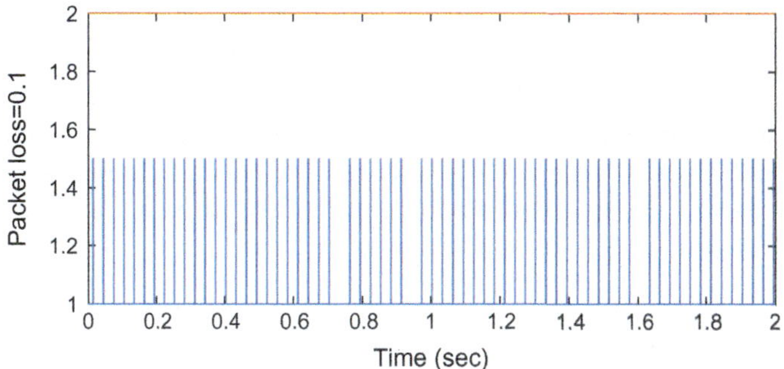

Fig. 4.35 Scheduling policy of sensor to controller when packet loss is 10%

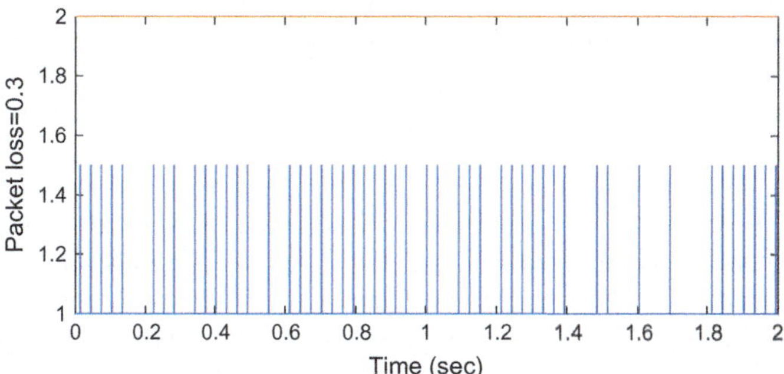

Fig. 4.36 Scheduling policy of sensor to controller when packet loss is 30%

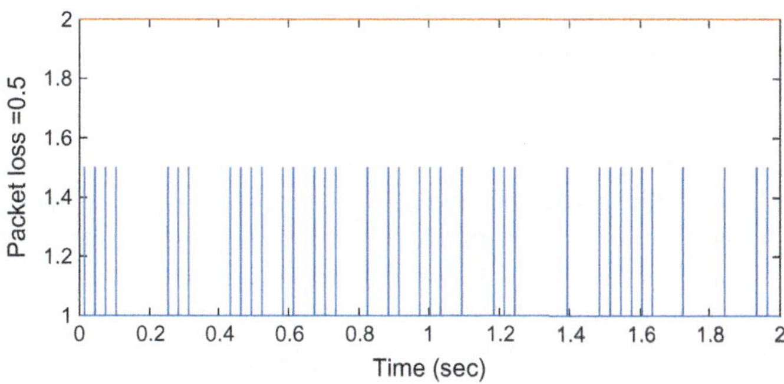

Fig. 4.37 Scheduling policy of sensor to controller when packet loss is 50%

controller delay. Apart from time delay compensation, the proposed algorithm was examined under packet loss condition. Figures 4.32, 4.33 and 4.34 show the results of tracking response under packet loss condition, while Figs. 4.35, 4.36 and 4.37 show the instances of packet drop. It can be observed from results that when the packet loss is 50%, the system goes to unstable condition. Thus, it can be concluded that the system shows the satisfactory response under 30% of packet loss for specified network delay with CAN as a communication medium.

4.7.2 Switched Ethernet as a Network Medium

In this section, the nature of the system is studied for Switched Ethernet type of communication medium considering the delays are deterministic in nature. In Switched Ethernet, the minimum frame size is 1024 bits and data transfer rate is 80000 bits/sec. So, the delay generated within the network to transfer the data packets from sensor to controller is $\tau_{sc} = 12.8$ ms and from controller to actuator is $\tau_{ca} = 12.8$ ms. The processing and the computational delays at sensor, controller and actuator are considered as 0.9, 0.5 and 0.5 ms, respectively. Thus, the total networked delay within the closed-loop system is computed as $\tau = 27.5$ ms. The fractional part of total network delay is obtained as $\tau' = 0.91$, $\tau'_{sc} = 0.426$ and $\tau'_{ca} = 0.426$ for sampling interval of $h = 30$ ms. The scheduling policies of sensor to controller and controller to actuator with network under ideal condition and bandwidth sharing condition are shown in Figs. 4.38 and 4.39, respectively. The trajectory response of the system for the network under ideal condition and traffic condition is shown in Figs. 4.40 and 4.41. It can be noticed that under both the situations, the output tracks the reference trajectory for the specified networked delay. In order to show the exact effect of time delay compensation in Switched Ethernet network at the output, results are magnified as shown in Figs. 4.42 and 4.43. It can be noticed that the effect of fractional delay from sensor to controller is compensated as the output tracks the reference input at $t = 14.7$ ms. The similar effect of time delay compensation for the network under traffic condition can be observed in sliding variable, Figs. 4.44 and 4.45, as well as control signal, Figs. 4.46 and 4.47. Observing the magnified results in Figs. 4.45 and 4.47, it can be noticed that the sliding variable and control signal are computed exactly after an interval of $t = 1.4$ ms even in the presence of sensor to controller delay. Apart from time delay compensation, the proposed algorithm was examined under packet loss condition. Figures 4.48, 4.49 and 4.50 show the results of tracking response under packet loss condition, while Figs. 4.51, 4.52 and 4.53 show the instances of packet drop. It can be observed from results that when the packet loss is 50%, the system goes to unstable condition. Thus, it can be concluded that the system shows the satisfactory response under 30% of packet loss for specified network delay with Switched Ethernet as a communication medium.

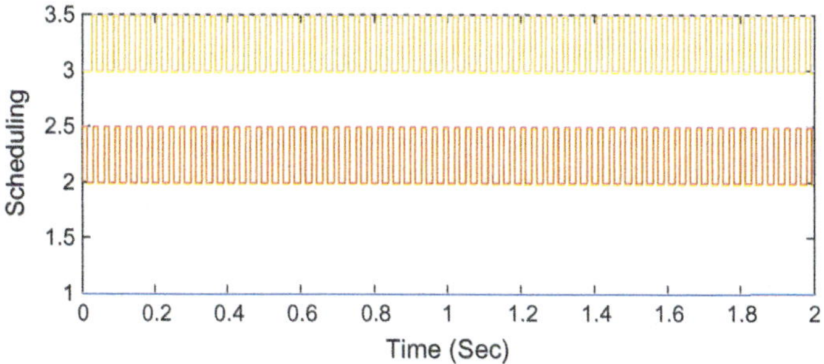

Fig. 4.38 Scheduling policies of sensor to actuator of Switched Ethernet network under ideal condition

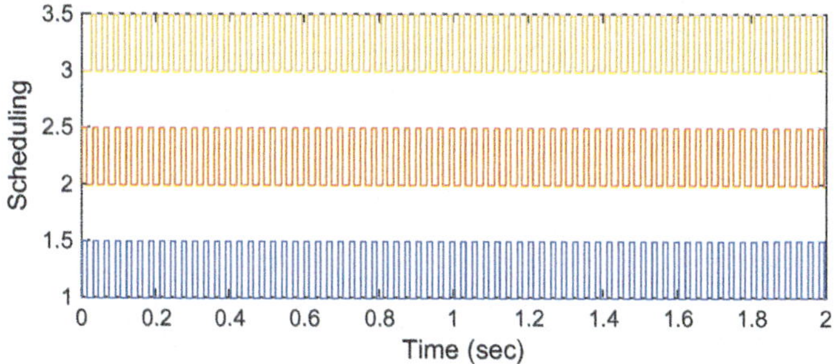

Fig. 4.39 Scheduling policies of sensor to actuator of Switched Ethernet network under traffic condition

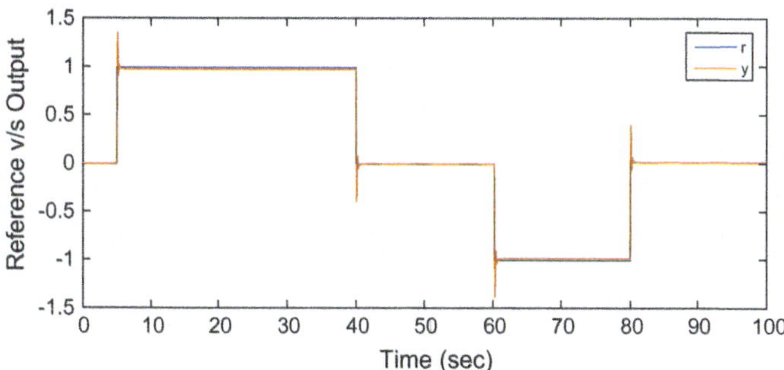

Fig. 4.40 Tracking response of the system with Switched Ethernet as networked medium under idle condition

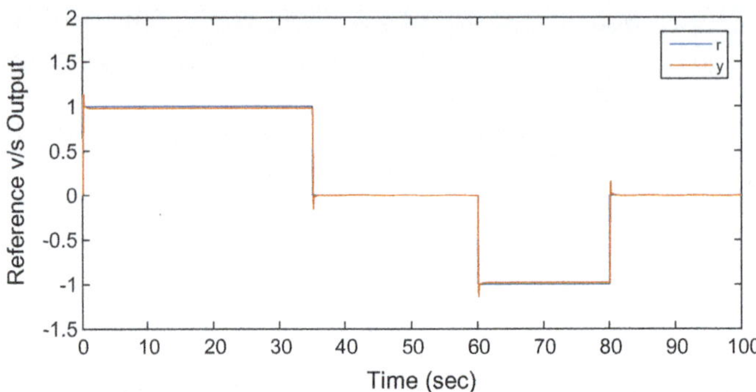

Fig. 4.41 Tracking response of the system with Switched Ethernet as networked medium with traffic condition

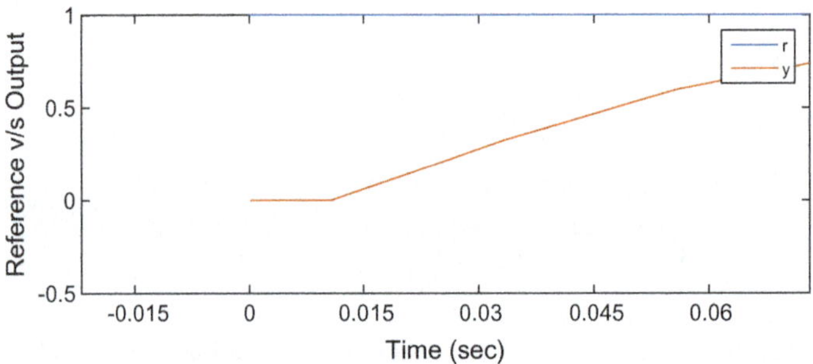

Fig. 4.42 Magnified tracking response of the system with network under ideal condition

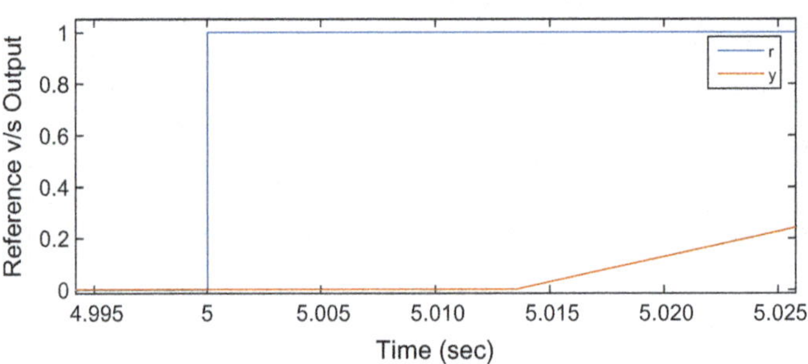

Fig. 4.43 Magnified tracking response of the system with network under traffic load

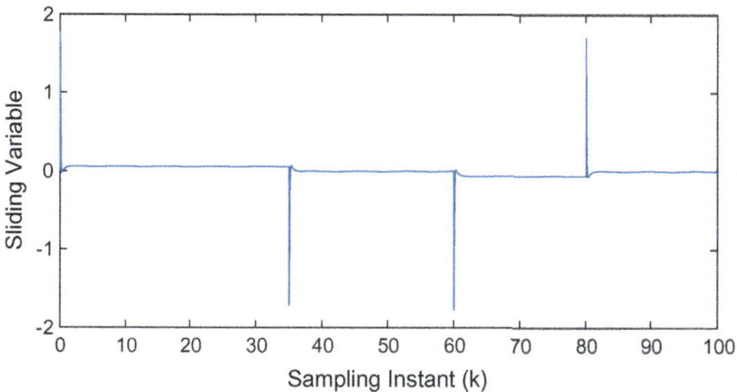

Fig. 4.44 Compensated sliding variable $s(k)$

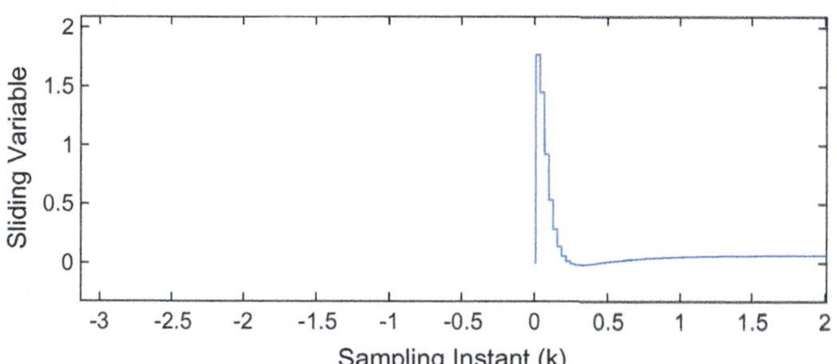

Fig. 4.45 Magnified compensated sliding variable $s(k)$

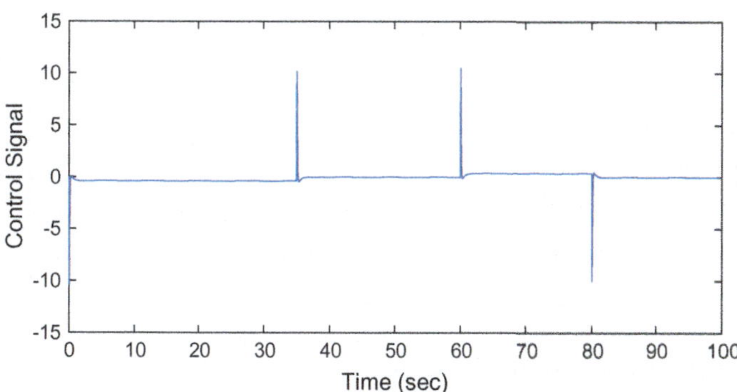

Fig. 4.46 Control signal $u(k)$

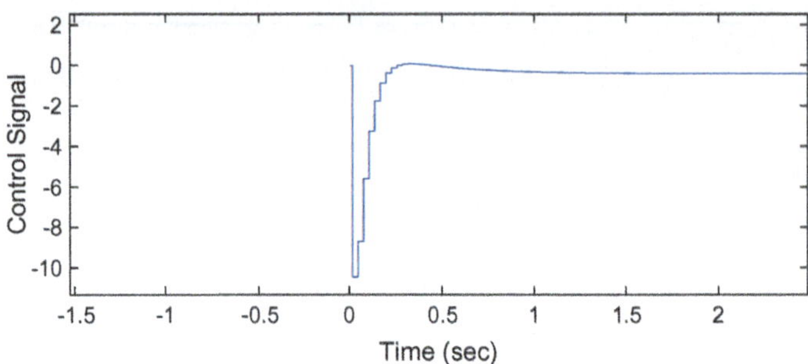

Fig. 4.47 Magnified control signal $u(k)$

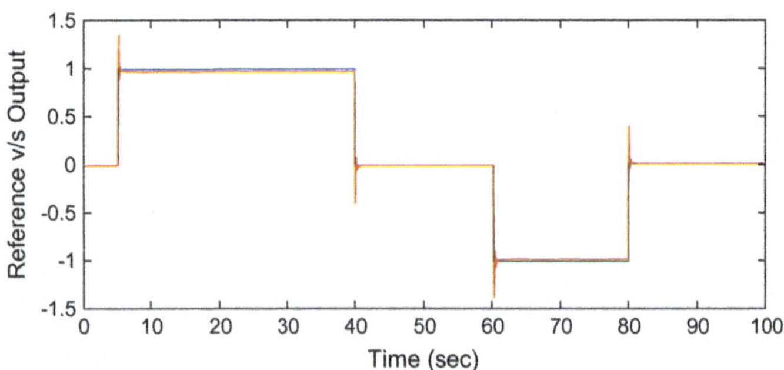

Fig. 4.48 Tracking response of system when packet loss is 10%

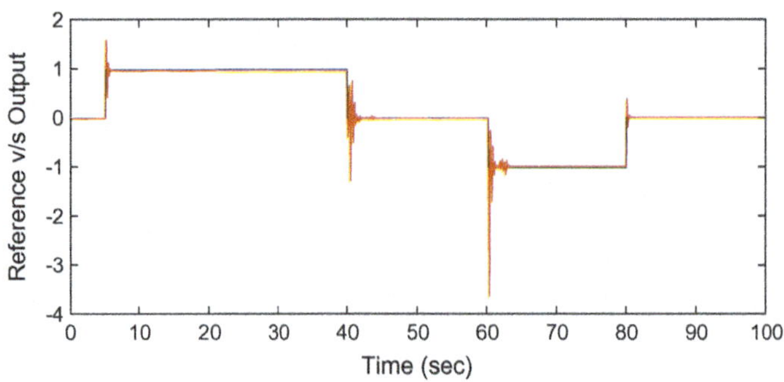

Fig. 4.49 Tracking response of system when packet loss is 30%

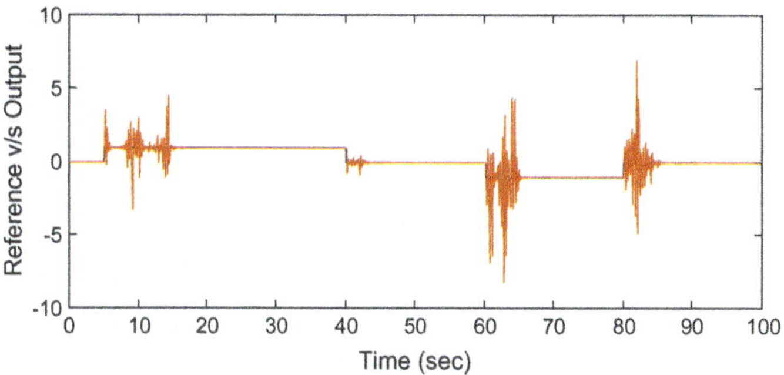

Fig. 4.50   Tracking response of system when packet loss is 50%

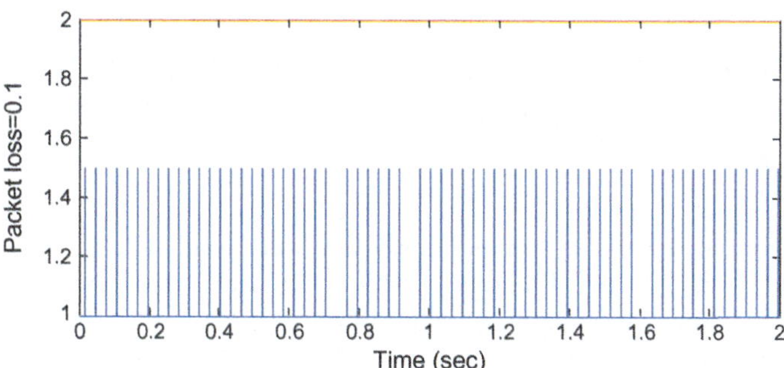

Fig. 4.51   Scheduling policy of sensor to controller when packet loss is 10%

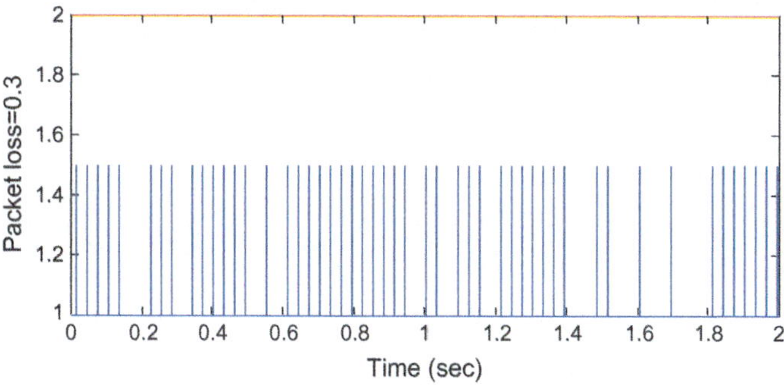

Fig. 4.52   Scheduling policy of sensor to controller when packet loss is 30%

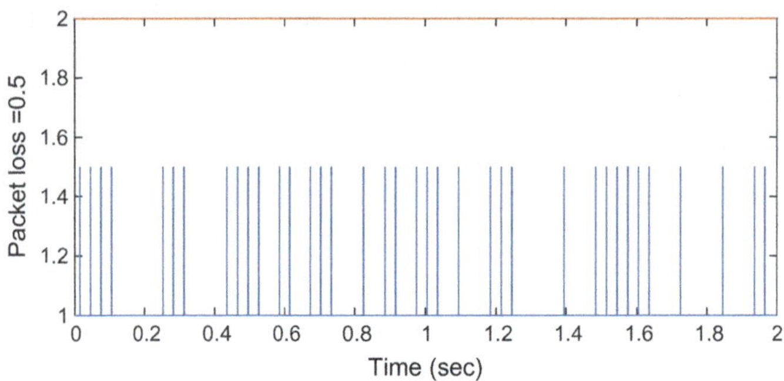

Fig. 4.53 Scheduling policy of sensor to controller when packet loss is 50%

4.7.3   Comparison of Proposed Algorithm with Conventional Sliding Mode Control Under CAN and Switched Ethernet as a Network Medium

In this section, the results of proposed algorithm were compared with conventional SMC for CAN and Switch Ethernet communication medium. Figures 4.54, 4.55, 4.56 and 4.57 show the comparative results of proposed algorithm and conventional SMC for discrete-time sliding mode control. It can be observed from comparison that conventional SMC shows unstable response for the specified networked delay. Thus, Thiran approximation proved to be efficient technique in discrete-time sliding mode control under packet loss condition and matched uncertainty. The comparison of proposed algorithm and conventional SMC for CAN and Switched Ethernet communication medium is summarized in Table 4.2.

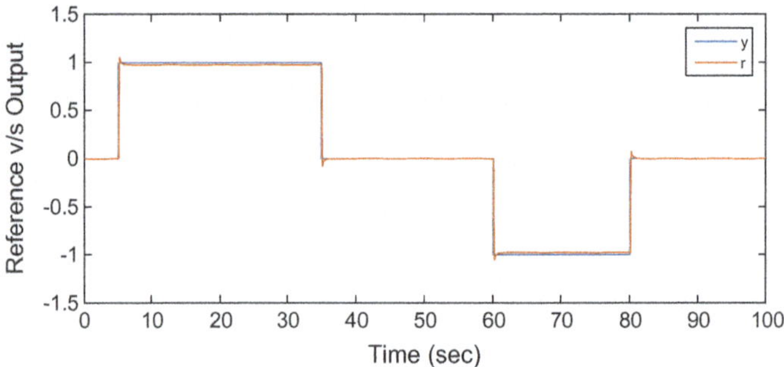

Fig. 4.54 Time delay compensation scheme with CAN as a networked medium

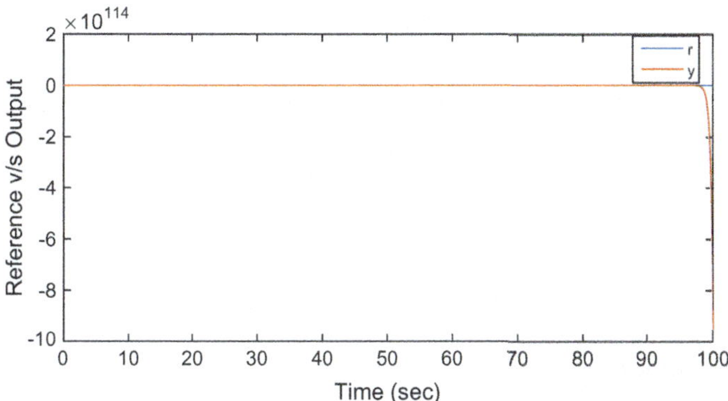

Fig. 4.55 Tracking response of conventional SMC with CAN as a communication medium

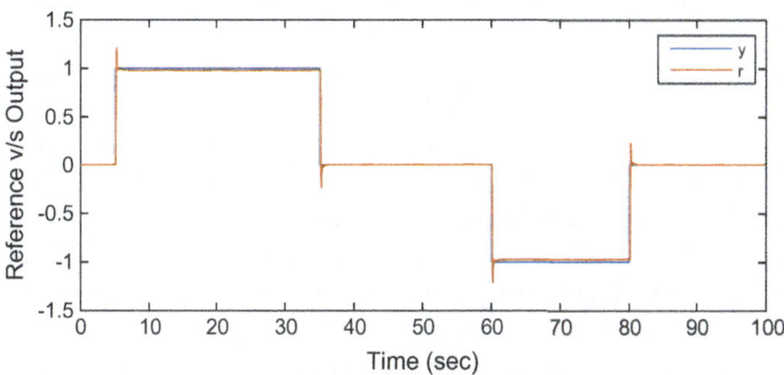

Fig. 4.56 Time delay compensation using Switched Ethernet as a communication medium

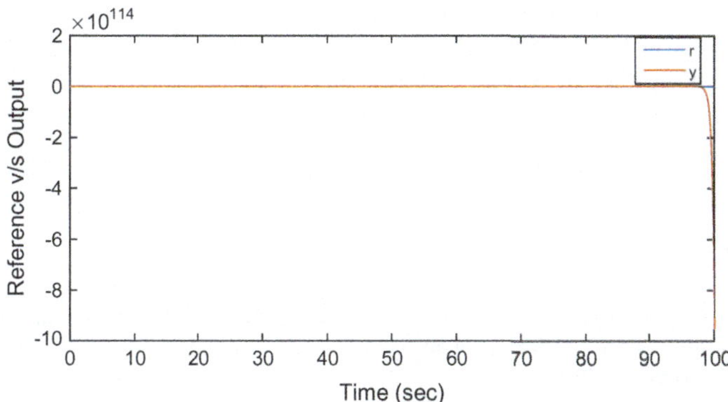

Fig. 4.57 Tracking response of conventional SMC with Switched Ethernet as a networked medium

Table 4.2 Comparison of proposed algorithm with conventional SMC in true time

Algorithm	Comparison results			
	τ_{CAN} (ms)	τ_{Ether} (ms)	T_s	Response
Conventional SMC	14.7	25.7	Undefined	Unstable
Proposed method	14.7	25.7	1 s	Stable

4.8 Conclusion

In this chapter, we explored Thiran's approximation technique for fractional delay compensation in discrete-time domain for designing non-switching type discrete-time sliding mode controller which computes the control actions in the presence of network delay and matched uncertainty. The stability of the closed-loop NCS is assured by using Lyapunov approach such that system states remain within the specified band. The effectiveness of the derived algorithms is tested using illustrative example and brushless DC motor set-up with deterministic networked delay and matched uncertainty. The experimental results are compared with switching SMC as well as conventional algorithm without delay compensation. The comparative results show that the non-switching type algorithm is most efficient technique than switching type algorithm. The simulation results and experimental results carried out for DC servo motor plant proved that the control algorithm designed using non-switching reaching law is robust and efficient algorithm as it provides the faster convergence with less chattering in discrete-time domain. Further, the efficiency of non-switching controller was examined under simulated CAN and Switched Ethernet networked medium using true time. The results show that the proposed control algorithm compensates the networked delay and performs well in the presence of network-induced delay.

References

1. D. Shah, A. Mehta, Discrete-time sliding mode controller subject to real-time fractional delays and packet losses for networked control system. Int. J. Control, Autom. Sys. (IJCAS) **15**(6), 2690–2703 (Dec. 2017)
2. D. Shah, A. Mehta, Fractional delay compensated discrete-time SMC for networked control system. Digital Commun. Networks (DCN), Elsevier, **2**(3), 385–390 (Dec. 2016)
3. A. Mehta, B. Bandyopadhyay, Multirate output feedback based stochastic sliding mode control. J. Dyn. Sys. Meas. Control **138**(12), 124503(1–6) (2016)
4. J. Thiran, Recursive digital filters with maximally flat group delay. IEEE Trans. Circ. Theory **18**(6), 659–664 (1971)
5. S. Eduardo, *Mathematical control theory: deterministic finite dimensional systems*, 2nd ed. (Springer, 1998) ISBN 0-387-98489-5

6. A. Bartoszewicz, P. Lesniewski, Reaching law approach to the sliding mode control of periodic review inventory systems. IEEE Trans. Autom. Sci. Eng. **11**(3), 810–817 (2014)
7. W. Gao, Y. Wang, A. Homaifa, Discrete-time variable structure control systems. IEEE Trans. Ind. Electron. **42**(2), 117–122 (1995)
8. J. Wu, T. Chen, Design of networked control systems with packet dropouts. IEEE Trans. Autom. Control **52**(7), 1314–1319 (2007)
9. D. Shah, A. Mehta, Design of robust controller for networked control system, in *Proceedings of IEEE International Conference on Computer, Communication and Control Technology* (2014) pp. 385–390
10. K. Astrom, J. Apkarian, P. Karam, M. Levis, J. Falcon, *Student Workbook: QNET DC Motor Control Trainer for NI ELVIS*, Quanser, 2015

Chapter 5
Multirate Output Feedback-Based Discrete-Time Sliding Mode Controller for NCS Having Deterministic Fractional Delay

Abstract In Networked Control System, the state feedback SMC is the simplest way for designing controller provided all state information is available. However, in many practical situations in network-based control system most of the states are observable but they are immeasurable. So it is essential to design the SMC controller using output information which is always available. This chapter presents the design of multirate output feedback-based discrete-time sliding mode controller in which the control input is computed using the system outputs and past control signals by taking full advantage of network transmission. A Thiran approximation technique is used to compensate the networked-induced delays for designing sliding mode controller (SMC). The stability of the closed-loop NCSs is derived using Lyapunov approach. Simulations results are presented to demonstrate the effectiveness of the proposed approach.

Keywords Multirate output feedback · Deterministic network delay
Thiran's approximation · Discrete-time sliding mode control · Stability

5.1 Problem Formulation

Consider the linear time-invariant SISO system with network delay as:

$$\dot{x}(t) = Ax(t) + Bu(t - \tau) + Dd(t), \qquad (5.1)$$

$$y(t) = Cx(t), \qquad (5.2)$$

where $x \in R^n$ is system state vector, $u \in R^m$ is control input, $y \in R^p$ is system output, $A \in R^{n \times n}$, $B \in R^{n \times m}$, $C \in R^{p \times n}$, $D \in R^{n \times m}$ are the matrices of appropriate dimensions, $d(t)$ is the matched bounded disturbance with $|d(t)| \le d_{max}$ and τ is the deterministic total networked-induced delay in continuous-time domain.

© Springer Nature Singapore Pte Ltd. 2018
D. H. Shah and A. Mehta, *Discrete-Time Sliding Mode Control for Networked Control System*, Studies in Systems, Decision and Control 132,
https://doi.org/10.1007/978-981-10-7536-0_5

The discrete form of system (5.1, 5.2) is:

$$x(k+1) = Fx(k) + Gu(k - \tau') + d(k), \tag{5.3}$$
$$y(k) = Cx(k), \tag{5.4}$$

where τ' is the deterministic total fractional network-induced delay in discrete-time domain, $F = e^{Ah}$, $G = \int_0^h e^{At} B dt$, $d(k) = \int_0^h e^{At} D d((k+1)h - t) dt \in O(h)$. Since $|d(t)| \leq d_{max}$, it can be inferred that $d(k)$ is also bounded and $O(h)$ [1]. For simplicity, it is assumed that $d(k)$ is slowly varying and remain constant over the interval $kh \leq t \leq (k+1)h$ [1].

The deterministic total fractional network-induced delay (τ') is denoted as,

$$\tau' = \frac{\tau}{h},$$

where h is the sampling interval.

Remark 10 It is considered that network-induced fractional delay (τ') in discrete time have non-integer values so it is required to compensate the delay at each sampling instants.

Assumption 5 The total network-induced delay (τ) is deterministic in nature satisfying,

$$\tau \prec h, \tag{5.5}$$

Remark 11 The above condition (5.5) indicates that the values of total fractional network-induced delay (τ') in discrete-time domain will be less than unity.

The total fractional network-induced delay (τ') is the combination of sensor to controller fractional delay (τ'_{sc}) and controller to actuator fractional delay (τ'_{ca}) which is given as,

$$\tau' = \tau'_{sc} + \tau'_{ca}, \tag{5.6}$$

where $\tau'_{sc} = \frac{\tau_{sc}}{h}$ and $\tau'_{ca} = \frac{\tau_{ca}}{h}$.

Assumption 6 The disturbance $d(k)$ is bounded by upper and lower bound as:

$$d_l \leq d(k) \leq d_u, \tag{5.7}$$

where d_l and d_u denote lower and upper bound of disturbance.

Remark 12 Without loss of generality, the sensor processing delay (τ_{sp}), controller computational delay (τ_{cp}) and actuator processing delay (τ_{ap}) are neglected as their values are negligible compared to network-induced delay (τ).

Now, we are ready to define the problem statement with above assumptions and conditions.

Problem Statement: The main objective is to design multirate output feedback-based discrete-time sliding mode control law for system (5.3, 5.4) in the presence of fractional delay τ'_{sc} and τ'_{ca} satisfying (5.5) and matched uncertainty satisfying (5.7).

The next section describes the mathematical derivation of sliding surface that compensates the effect of fractional delay (τ') in the presence of matched uncertainty $d(k)$ using multirate output feedback approach.

5.2 Sliding Surface Using Multirate Output Feedback

The schematic diagram of NCS with time delay compensator and multirate output feedback (MROF) approach is depicted in Fig. 5.1. As shown, the state information and control information are exchanged between plant and controller through the networks suffering from sensor to controller delay (τ_{sc}) and controller to actuator delay (τ_{ca}). The state information is computed based on multirate output feedback approach. In this approach, the system states are estimated through the output information and error between the computed state and actual state of the system goes to zero once the multirate output sample is available [2–4]. While in conventional output feedback the error between the actual states and estimated states decreases asymptotically and approaches to zero at infinite time. In multirate output feedback technique the control inputs and plant output signals are sampled at different sampling intervals. It is assumed that the sensor output measurements are faster than control input signals. The mathematical expression of control input signal sampled at τ_c sampling interval is given by:

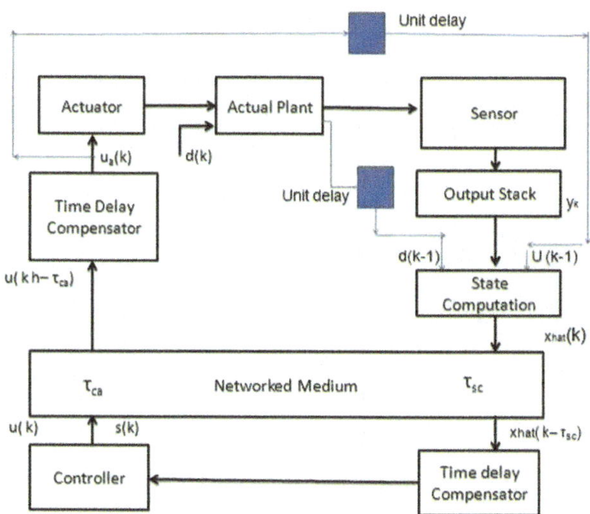

Fig. 5.1 Block diagram of NCS with multirate output feedback approach

$$x((k+1)\tau_c) = F_{\tau_c} x(k\tau_c) + G_{\tau_c} u(k\tau_c) + d(k\tau_c), \qquad (5.8)$$

$$y(k\tau_c) = Cx(k\tau_c). \qquad (5.9)$$

Let the output samples be sampled at ζ sampling interval given by,

$$\zeta = \frac{\tau_c}{\Lambda}, \qquad (5.10)$$

where Λ is the positive integer $\geq$ observability index of the system.

Similarly, the mathematical expression of the sensor output data sampled at ζ sampling interval is given as:

$$x((k+1)\zeta) = F_\zeta x(k\zeta) + G_\zeta u(k\zeta) + d(k\zeta), \qquad (5.11)$$

$$y(k\zeta) = Cx(k\zeta). \qquad (5.12)$$

Remark 13 The stability of closed-loop system is guaranteed, if pair (F_{τ_c}, G_{τ_c}) is controllable and pair (F_ζ, C) is observable.

The output of multirate output feedback [5] is expressed in mathematical form as:

$$\hat{x}(k) = H_y y_k + H_u u(k-1) + H_d d(k-1), \qquad (5.13)$$

where y_k represents the output stack, $u(k-1)$ represents the past control input signal, $d(k-1)$ represents the past disturbance signal which are available from measurements.

$$H_y = F_{\tau_c}(Z_0^T Z_0)^{-1} Z_0^T, \ H_u = G_{\tau_c} - H_y X_0, \ H_d = d_{\tau_c} - H_y C_d$$

$$Z_0 = \begin{bmatrix} C \\ CA_\zeta \\ \vdots \\ \vdots \\ CA_\zeta^{N-1} \end{bmatrix}, \ X_0 = \begin{bmatrix} 0 \\ CB_\zeta \\ \vdots \\ \vdots \\ C\Sigma_{i=0}^{N-1}(A_\zeta)^i B_\zeta \end{bmatrix},$$

$$C_d = \begin{bmatrix} 0 \\ C_d \zeta \\ \vdots \\ \vdots \\ C\Sigma_{i=0}^{N-1}(A_\zeta)^i B_\zeta \end{bmatrix}, \ y_k = \begin{bmatrix} y((k-1)\tau_c) \\ y((k-1)\tau_c + \zeta) \\ \vdots \\ \vdots \\ y(k\tau_c - \zeta) \end{bmatrix}$$

Thus, observing Eq. (5.13) it can be noticed that the output of multirate depends on past output samples, past control signal and past disturbance signal. Thus, the output of sensor computed based on multirate output feedback approach will be applied to the controller through network.

For designing DTSMC, we propose the sliding surface using Thiran's approximation rule and multirate output feedback approach in the form of Lemma 3 as under.

Lemma 3 *The sliding surface for the given system* (5.3, 5.4) *with sensor to controller fractional network delay satisfying Assumptions 5 and 6 are given as:*

$$s(k) = C_s \hat{x}(k) - \alpha C_s \hat{x}(k-1),\tag{5.14}$$

where
$\alpha = \frac{\hat{\tau}_{sc}}{1+\hat{\tau}_{sc}}$, C_s *represents the sliding gain.*

Proof Let the sliding surface with network fractional delay from sensor to controller (τ'_{sc}) is given by:

$$s(k) = C_s \hat{x}(k - \tau'_{sc}),\tag{5.15}$$

where C_s indicates the sliding gain, $\hat{x}(k - \hat{\tau}_{sc})$ indicates the delayed state vector. The value of sliding gain C_s is computed using discrete LQR method with proper selection of Q and R matrices [6].

Applying z-transform to Eq. (5.15) we get:

$$s(z) = C_s \hat{x}(z) z^{-\tau'_{sc}},\tag{5.16}$$

where $\tau'_{sc} = v = \tau_{sc}/h$.

Using Thiran's approximation [7], $z^{-\tau'_{sc}}$ can be approximated as,

$$z^{-\tau'_{sc}} = \Sigma_{k=0}^{1}(-1)^k \binom{l}{k} \Pi_{i=0}^{1} \frac{2\tau'_{sc}+i}{2\tau'_{sc}+k+i} z^{-k}.\tag{5.17}$$

The above Eq. (5.17) can be further expanded as:

$$z^{-\tau'_{sc}} = \left[(-1)^0 \binom{1}{0} \left\{ \frac{2\tau'_{sc}}{2\tau'_{sc}} * \frac{2\tau'_{sc}+1}{2\tau'_{sc}+1} \right\} z^0 + (-1)^1 \binom{1}{1} \right.$$
$$\left. \left\{ \frac{2\tau'_{sc}}{2\tau'_{sc}+1} * \frac{2\tau'_{sc}+1}{2\tau'_{sc}+2} \right\} z^{-1} \right].\tag{5.18}$$

On simplifying we get,

$$z^{-\tau'_{sc}} = 1 - \alpha z^{-1}.\tag{5.19}$$

where $\alpha = \frac{\tau'_{sc}}{\tau'_{sc}+1}$.

Substituting Eqs. (5.19) into (5.16),

$$s(z) = [1 - \alpha z^{-1}]\hat{x}(z) C_s.\tag{5.20}$$

Further solving,

$$s(z) = \hat{x}(z)C_s - \alpha z^{-1}\hat{x}(z)C_s. \tag{5.21}$$

Applying inverse z-transform we have,

$$s(k) = C_s\hat{x}(k) - \alpha C_s\hat{x}(k-1). \tag{5.22}$$

This completes the *proof*.

Observing Eq. (5.22), it can be noticed that the sliding surface $s(k)$ depends past output samples, past control signal, past disturbance signal and parameter "α" approximated through Thiran's approximation rule. Thus, the effect of sensor to controller fractional delay τ'_{sc} in the sliding surface at each sampling instants h is compensated through proposed technique.

Now, we are ready to propose multirate output feedback-based discrete-time sliding mode control law using the sliding surface (5.14).

5.3 Design of Multirate Output Feedback Discrete-Time Networked Sliding Mode Control

In this section, discrete-time sliding mode control law is derived for deterministic network fractional delays (τ'_{sc} and τ'_{ca}) along with its stability using sliding surface (5.14) represented in form of Theorem 5.1.

Theorem 5.1 *The discrete-time sliding surface (5.14) is reached in a finite time in the presence of networked delays satisfying (5.5) and matched uncertainty (5.7) provided the control law is designed as:*

$$u(k) = -(C_sG)^{-1}[H_{dy}y_k + H_{ud}u(k-1) + H_{dd}d(k-1) \tag{5.23}$$
$$- Js(k) - d_s(k) + d_1] - d(k).$$

where
$H = (C_sF), \ I = \alpha C_s, \ J = [1 - q(s(k))], \ (H-I)H_y = H_{dy} \ (H-I)H_u = H_{ud}, \ (H-I)H_d = H_{dd}.$

Proof The reaching law proposed in [8] is used to derive the control law since it provides faster convergence. The reaching law in the presence of sensor to controller fractional delay (τ'_{sc}) is given by:

$$s[(k+1)] = \{1 - q[s(k)]\}s(k) - d(k) + d_1, \tag{5.24}$$

where
$\{q[s(k)]\} = \frac{\psi}{\psi+|s(k)|}$, $d(k)$ represents the disturbance,

$d_1 = \frac{d_u + d_l}{2}$, mean value of $d(k)$, $d_2 = \frac{d_u - d_l}{2}$, deviated value of $d(k)$ and ψ is the designer's constant satisfying,

$$\psi \geq d_2. \tag{5.25}$$

The reaching law in Eq. (5.24) contains the disturbance term which is applied through the network. Thus, the compensated disturbance $d_s(k)$ is given as:

$$d_s(k) = d(k) - \alpha d(k - 1). \tag{5.26}$$

Using Eqs. (5.14, 5.24) can be rewritten as

$$\hat{x}(k + 1)C_s - \alpha C_s \hat{x}(k) = [1 - q(s(k))]s(k) - d_s(k) + d_1. \tag{5.27}$$

Remark 14 It is noticed from Eq. (5.22) that the sensor to controller fractional delay is compensated at the sliding surface while controller to actuator fractional delay is compensated at the actuator side. So, the effect of delay will not be observed at the controller side for computing the control signal as well as actuator side. Thus without loss of generality, the control signal in Eq. (5.3) is given as

$$u(k - \tau') = u(k) \tag{5.28}$$

Substituting the value of $\hat{x}(k + 1)$ in Eq. (5.27),

$$C_s[F\hat{x}(k) + Gu(k) + d(k)] - \alpha C_s \hat{x}(k) = [1 - q(s(k))]s(k) - d_s(k) + d_1. \tag{5.29}$$

Further simplification gives

$$C_s F\hat{x}(k) + C_s Gu(k) + C_s d(k) - \alpha C_s \hat{x}(k) = [1 - q(s(k))]s(k) - d_s(k) + d_1. \tag{5.30}$$

The above Eq. (5.30) further can be expressed in the terms of control law as,

$$\begin{aligned}
u(k) = &- (C_s G)^{-1}[H_{dy}y_k + H_{ud}u(k - 1) \\
&+ H_{dd}d(k - 1) - Js(k) - d_s(k) + d_1] - d(k).
\end{aligned} \tag{5.31}$$

This completes the *proof*.

Observing the Eq. (5.31), it can be noticed that the control law is computed based on the information available in the output stack y_k and not the state information. Thus, using sliding surface (5.14) and control law (5.23) the stability of the closed-loop NCS is derived such that the system states will remain within the band in the presence of network fractional delay (τ') and matched uncertainty $d(k)$.

5.4 Stability

Theorem 5.2 *The system states in (5.3, 5.4) will slides on the sliding surface (5.14) and maintain on it with the controller designed in (5.23) under network fractional delay (τ') satisfying (5.5) and matched uncertainty (5.7) such that for any $\psi \geq d_2$ and $\gamma \prec 0$ the following condition should hold true:*

$$0 \preceq \kappa \prec s^T(k)s(k). \tag{5.32}$$

where
$$\kappa = [[1 - q(s(k))]s(k) - d_c(k) + d_1]^T * [[1 - q(s(k))]s(k) - d_c(k) + d_1]$$

Proof Consider the sliding surface (5.14) as,

$$s(k+1) = C_s \hat{x}(k+1) - \alpha C_s \hat{x}(k). \tag{5.33}$$

Let the Lyapunov function be given by,

$$V_s(k) = s^T(k)s(k). \tag{5.34}$$

Taking forward difference of the above Eq. (5.34),

$$\Delta V_s(k) = s^T(k+1)s(k+1) - s^T(k)s(k). \tag{5.35}$$

Substituting the value of $s(k+1)$ from Eq. (5.14) we get,

$$\Delta V_s(k) = [C_s \hat{x}(k+1) - \alpha C_s \hat{x}(k)]^T [C_s \hat{x}(k+1) - \alpha C_s \hat{x}(k)] \\ - s^T(k)s(k). \tag{5.36}$$

Substituting the value of $\hat{x}(k+1)$ we get,

$$\Delta V_s(k) = [C_s[F\hat{x}(k) + G(u(k) + d(k))] - \alpha C_s \hat{x}(k)]^T [C_s[F\hat{x}(k) + G(u(k) \\ + d(k))] - \alpha C_s \hat{x}(k)] - s^T(k)s(k). \tag{5.37}$$

Substituting the value of $u(k)$ and further solving it we have,

$$\Delta V_s(k) = [[1 - q(s(k))]s(k) - d_c(k) + d_1]^T * [[1 - q(s(k))]s(k) - d_c(k) \\ + d_1] - s^T(k)s(k). \tag{5.38}$$

It can be seen that the term $[[1 - q(s(k))]s(k) - d_c(k) + d_1]$ contains the disturbance term in $\Delta V_s(k)$. Let $\kappa = [[1 - q(s(k))]s(k) - d_c(k) + d_1]^T * [[1 - q(s(k))]s(k) - d_c(k) + d_1]$

Then we have,

$$\Delta V_s(k) = \kappa - s^T(k)s(k). \tag{5.39}$$

If κ is closed tuned to zero by appropriately selecting the parameter ψ, then $s^T(k)s(k)$ will be larger than κ. Therefore, for any constant parameter γ, we have $\Delta V_s(k) \prec -\gamma s^T(k)s(k)$ which guarantees the convergence of $\Delta V_s(k)$.
This completes the *proof*.

5.5 Results and Discussion

In this section, the efficacy of the designed control algorithm based on MROF approach is validated in the presence of fractional network delay and matched uncertainty applied at the input channel of the system. The simulation results are carried out using illustrative example [9] in MATLAB environment considering different fractional delays.

5.5.1 Simulation Results

Consider the continuous-time LTI system as,

$$\dot{x}(t) = Ax(t) + Bu(t - \tau) + Dd(t), \tag{5.40}$$
$$y(t) = Cx(t), \tag{5.41}$$

where
$$A = \begin{bmatrix} -0.7 & 2 \\ 0 & -1.5 \end{bmatrix}, B = \begin{bmatrix} -0.03 \\ -1 \end{bmatrix},$$
$$C = \begin{bmatrix} 1 & 0 \end{bmatrix}, D = \begin{bmatrix} 1 \\ 1 \end{bmatrix}, d(t) = 0.2\sin(0.086t).$$

Discretizing the above system with sampling interval of $h = 30\,\text{ms}$ we get,

$$x(k + 1) = Fx(k) + Gu(k - \tau') + d(k), \tag{5.42}$$
$$y(k) = Cx(k), \tag{5.43}$$

where
$$F = \begin{bmatrix} 0.9792 & 0.05805 \\ 0 & 0.956 \end{bmatrix}, G = \begin{bmatrix} -0.001771 \\ -0.02934 \end{bmatrix},$$
$$C = \begin{bmatrix} 1 & 0 \end{bmatrix}.$$

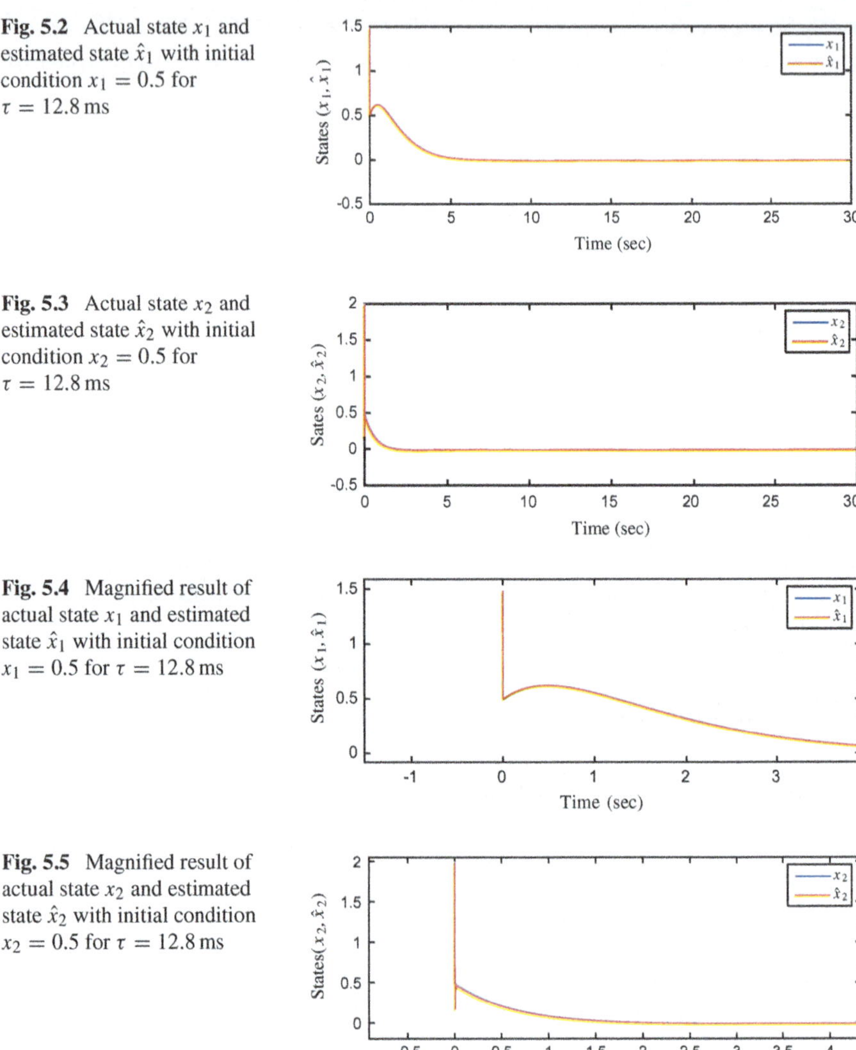

Fig. 5.2 Actual state x_1 and estimated state $\hat{x}_1$ with initial condition $x_1 = 0.5$ for $\tau = 12.8\,\text{ms}$

Fig. 5.3 Actual state x_2 and estimated state $\hat{x}_2$ with initial condition $x_2 = 0.5$ for $\tau = 12.8\,\text{ms}$

Fig. 5.4 Magnified result of actual state x_1 and estimated state $\hat{x}_1$ with initial condition $x_1 = 0.5$ for $\tau = 12.8\,\text{ms}$

Fig. 5.5 Magnified result of actual state x_2 and estimated state $\hat{x}_2$ with initial condition $x_2 = 0.5$ for $\tau = 12.8\,\text{ms}$

Figures 5.2, 5.3, 5.4, 5.5, 5.6, 5.7, 5.8, 5.9, 5.10, 5.11, 5.12, 5.13, 5.14, 5.15, 5.16, 5.17 and 5.18 show the nature of system using multirate output feedback approach under networked environment. The robustness of the system is validated by applying slow time-varying disturbance applied at the input side of channel. The networked delay is considered as the time required for the data packets to travel from sensor to controller and controller to actuator. The sliding gain parameter is calculated using discrete LQR method which comes out to be $C_s = [-1.77 \ -2.766]$ for $Q = diag(1000, 1000)$ and $R = 1$. The sliding band $|s(k)|$ remains same as that of previous cases. The initial condition of the system in both the cases are considered as $[x_1 \ x_2] = [0.5 \ 0.5]^T$.

Fig. 5.6 Sliding surface $s(k)$ for $\tau = 12.8\,\text{ms}$

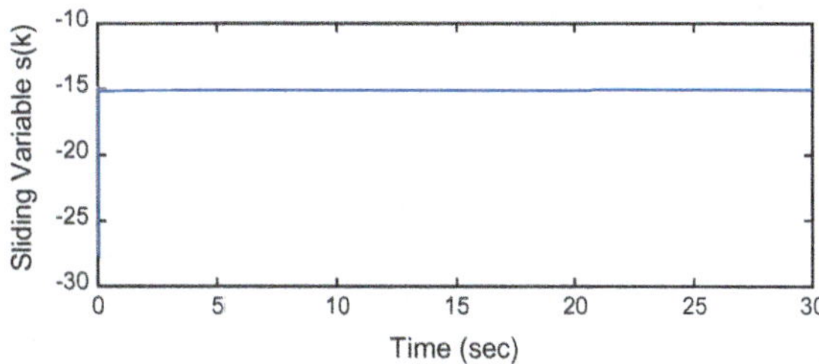

Fig. 5.7 Magnified sliding surface $s(k)$ for $\tau = 12.8\,\text{ms}$

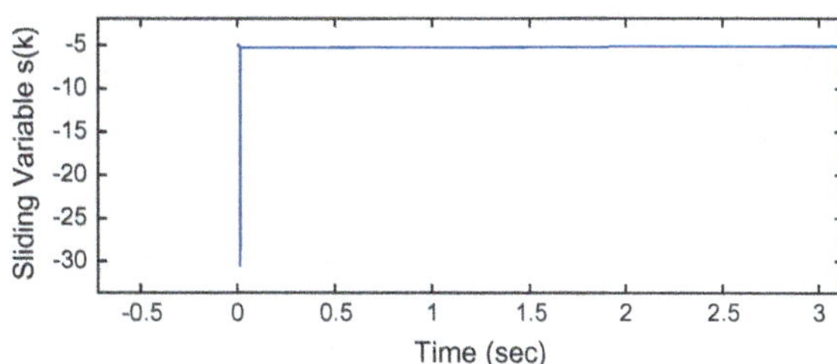

Fig. 5.8 Control signal $u(k)$ for $\tau = 12.8\,\text{ms}$

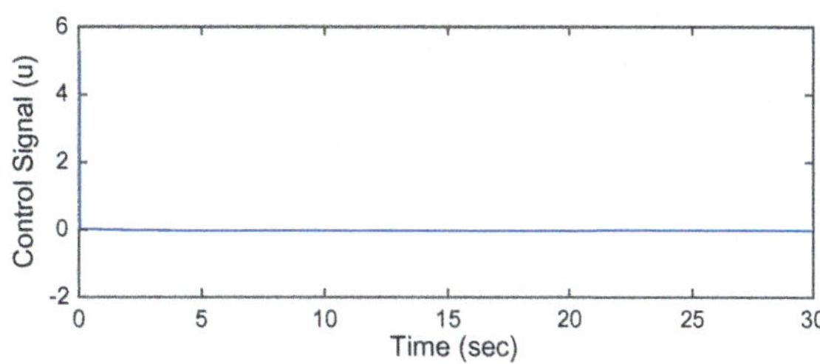

Fig. 5.9 Magnified control signal $u(k)$ for $\tau = 12.8\,\text{ms}$

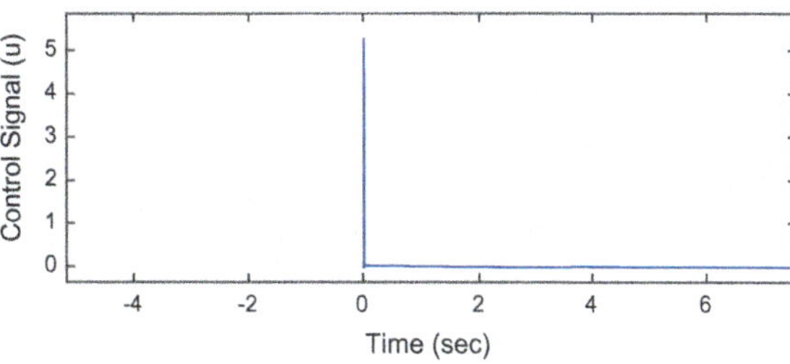

Fig. 5.10 Actual state x_1 and estimated state $\hat{x}_1$ with initial condition $x_1 = 0.5$ for $\tau = 25.6\,\text{ms}$

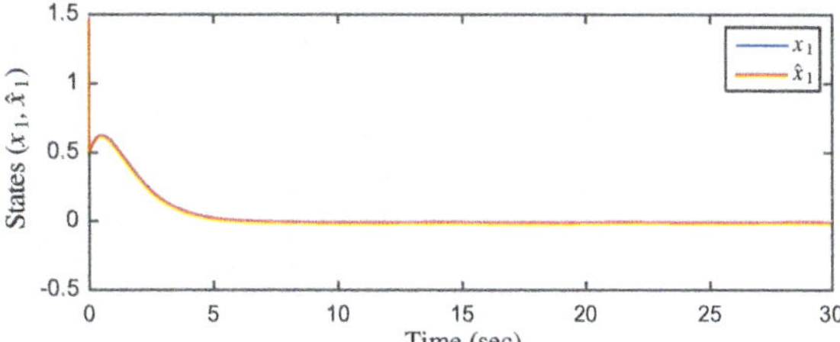

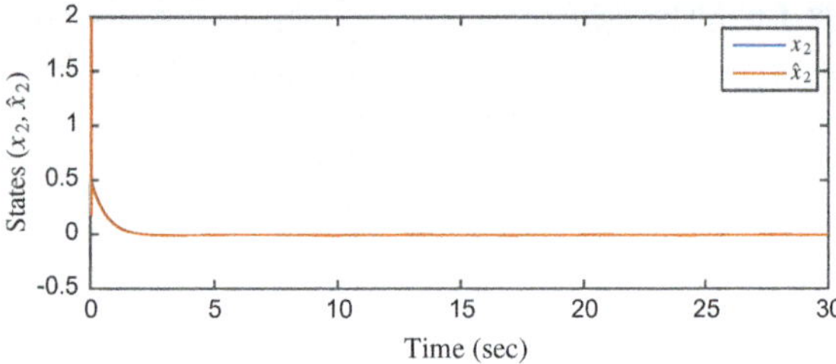

Fig. 5.11 Actual state x_2 and estimated state $\hat{x}_2$ with initial condition $x_2 = 0.5$ for $\tau = 25.6\,\text{ms}$

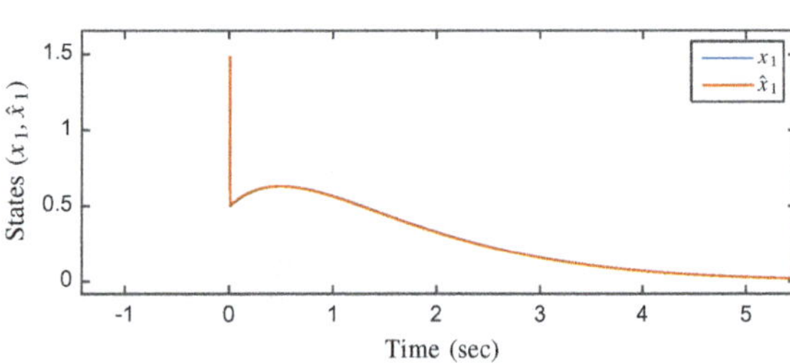

Fig. 5.12 Magnified result of actual state x_1 and estimated state $\hat{x}_1$ with initial condition $x_1 = 0.5$ for $\tau = 25.6\,\text{ms}$

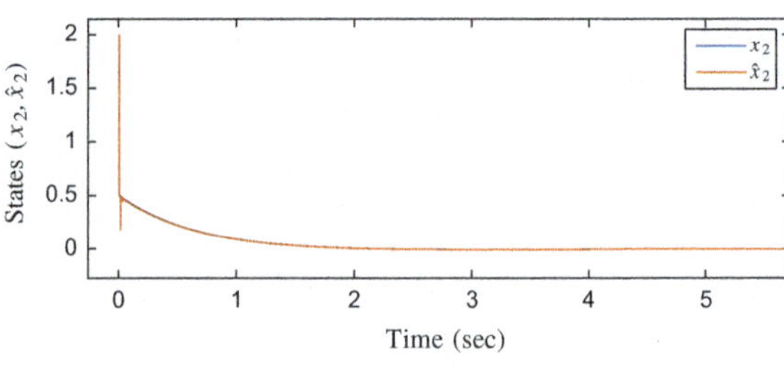

Fig. 5.13 Magnified result of actual state x_2 and estimated state $\hat{x}_2$ with initial condition $x_2 = 0.5$ for $\tau = 25.6\,\text{ms}$

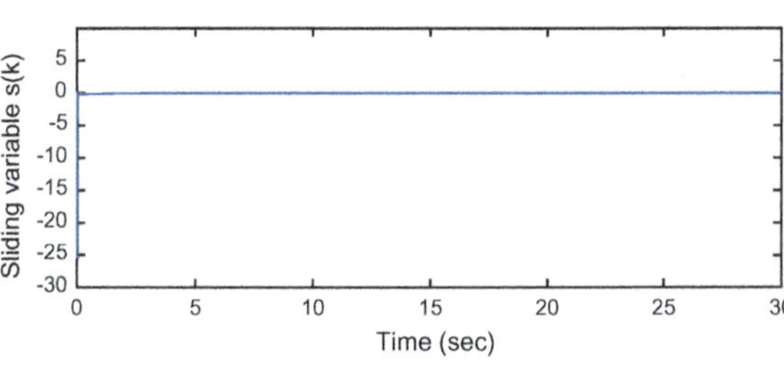

Fig. 5.14 Sliding surface $s(k)$ for $\tau = 25.6\,\text{ms}$

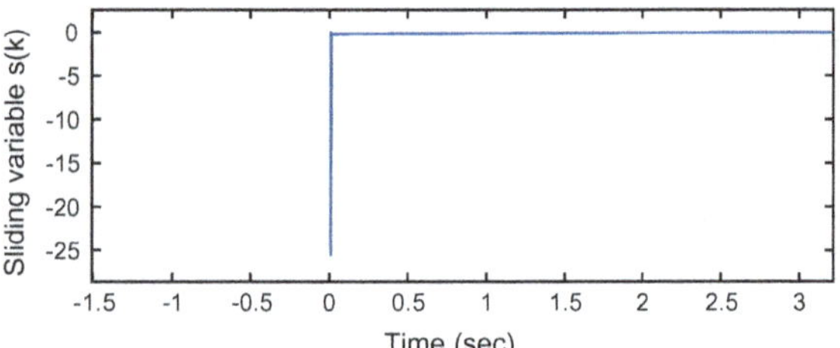

Fig. 5.15 Magnified sliding surface $s(k)$ for $\tau = 25.6\,\text{ms}$

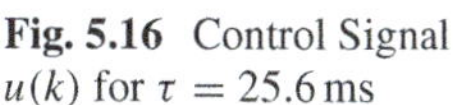

Fig. 5.16 Control Signal $u(k)$ for $\tau = 25.6\,\text{ms}$

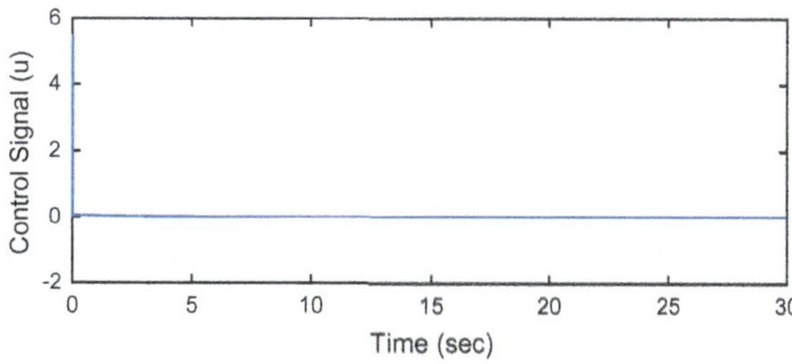

Fig. 5.17 Magnified control signal $u(k)$ for $\tau = 25.6\,\text{ms}$

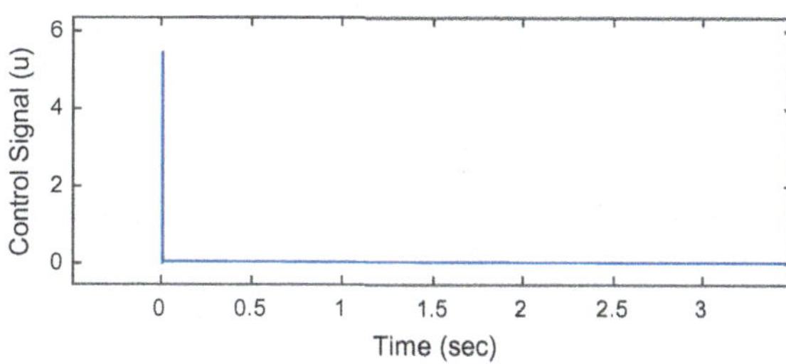

Fig. 5.18 Output stack y_k

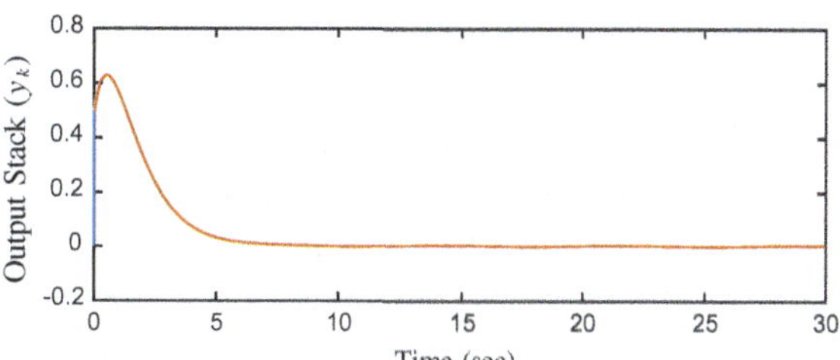

The sensor output signal is sampled at $\zeta = 2\,\text{ms}$ and the control input signal is sampled at $\tau_c = 6\,\text{ms}$ considering $\Lambda = 3$. The efficacy of the proposed algorithm is checked under networked delay of $\tau = 12.8$ and $\tau = 25.6\,\text{ms}$, respectively.

Figures 5.2, 5.3, 5.4, 5.5, 5.6, 5.7, 5.8 and 5.9 show the regulatory response of state variables, sliding variable and control signal for $\tau = 12.8\,\text{ms}$ with $\tau_{sc} = 6.4\,\text{ms}$ and $\tau_{ca} = 6.4\,\text{ms}$. Since the system is discretized at $h = 30\,\text{ms}$, the fractional part of delay is computed to be $\tau' = 0.426$, $\tau'_{sc} = 0.213$ and $\tau'_{ca} = 0.213$, respectively. Figures 5.2 and 5.3 show the regulatory response of the system states for the specified networked delay. It can be observed that the actual states (x_1, x_2) and estimated states $(\hat{x}_1, \hat{x}_2)$ both converges to zero from their given initial condition. In order to prove the effect of multirate output feedback and time delay compensation, the results are magnified as shown in Figs. 5.4 and 5.5, respectively. It can be noticed that in both cases the estimated state variables follow the actual state variable at $\zeta = 2\,\text{ms}$ which means the error becomes zero once the multirate output sample is available. Apart from these, both the states are computed from first sampling instants. Thus, the effect of fractional time delay from sensor to controller and controller to actuator is compensated. The similar effect of compensation is observed in sliding variable and control signal results shown in Figs. 5.6, 5.7, 5.8 and 5.9. Observing the magnified results shown in Figs. 5.7 and 5.9, it can be noticed that the sliding variable and control signal both are computed from first sampling instants even in the presence of sensor to controller fractional delay.

Figures 5.10, 5.11, 5.12, 5.13, 5.14, 5.15, 5.16 and 5.17 show the response of the system in terms of state variables, sliding variable and control signal for $\tau = 25.6$ ms with $\tau_{sc} = 12.8$ ms and $\tau_{ca} = 12.8$ ms. The fractional part of delay is computed to be $\tau' = 0.8533$, $\tau'_{sc} = 0.426$ and $\tau'_{ca} = 0.426$ for $h = 30$ ms. Figures 5.10 and 5.11 show the regulatory response of the system state variables for specified networked delay. It can be noticed that actual states (x_1, x_2) and estimated states $(\hat{x}_1, \hat{x}_2)$ both converges to zero from their given initial condition. The magnified regulatory response of the same is shown in Figs. 5.12 and 5.13. It can be observed that both state variables start converging to zero at first sampling instant and the estimated states follow the actual state variables at $\zeta = 2$ ms. Thus, the effect of network delay from sensor to controller and controller to actuator is nullified at the output. The same effect is observed in sliding variable Fig. 5.14 as well as control signal Fig. 5.16. Observing the magnified results shown in Figs. 5.15 and 5.17, it can be concluded that the effect of fractional delay from sensor to controller is compensated through Thiran's approximation. Figure 5.18 shows the result of output stack for both the cases.

Thus, from above results it can be noticed that the proposed control algorithm works efficiently for the networked delay of 12.8–25.6 ms. The sliding surface and control algorithm derived using multirate output feedback approach with Thiran's approximation takes care of sensor to controller fractional delay and controller to actuator fractional delay in the presence of matched uncertainty.

5.6 Conclusion

In this chapter, idea of compensating the fractional delay in forward channel is introduced. The sensor to controller fractional delay is compensated using Thiran's approximation at the sliding surface. A multirate output feedback technique is used to estimate the states at the plant side. Using this novel approach, a multirate output feedback discrete-time networked sliding mode control law is derived that compute the control sequences in the presence of network fractional delay and matched uncertainty. The main advantage of using multirate output feedback approach is that the system states are estimated exactly after one sampling instant. Stability of the closed-loop NCS is ensured using Lyapunov approach such that system states would remain within the band under network non-idealities. The simulation results carried out using proposed technique shows the efficacy of the algorithm in discrete time for deterministic type of network fractional delay.

References

1. A. Mehta, B. Bandyopadhyay, Multirate output feedback based stochastic sliding mode control. J. Dyn. Syst. Measur. Control **138**(12), 124503(1–6) (2016)
2. S. Janardhanan, B. Bandyopadhyay, Output feedback sliding mode control for uncertain system using fast output sampling technique. IEEE Trans. Ind. Electron. **53**, 1677–1682 (2006)
3. S. Janardhanan, V. Kariwala, Multirate-output-feedback-based LQ-optimal discrete-time sliding mode control. IEEE Trans. Autom. Control **53**(1), 367–373 (2008)
4. A.J. Mehta, B. Bandyopadhyay, in *Frequency-Shaped and Observer-Based Discrete-Time Sliding Mode Control*. SpringerBriefs in Applied Sciences and Technology (Springer, New Delhi, 2015)
5. D. Shah, A. Mehta, Output feedback discrete-time networked sliding mode control, in *IEEE Proceedings of Recent Advances in Sliding Modes (RASM)* (2015), pp. 1–7
6. S. Eduardo *Mathematical Control Theory: Deterministic Finite Dimensional Systems*, 2nd edn. (Springer, New York, 1998), ISBN 0-387-98489-5
7. J. Thiran, Recursive digital filters with maximally flat group delay. IEEE Trans. Circuit Theory **18**(6), 659–664 (1971)
8. A. Bartoszewicz, P. Lesniewski, Reaching law approach to the sliding mode control of periodic review inventory systems. IEEE Trans. Autom. Sci. Eng. **11**(3), 810–817 (2014)
9. D. Shah, A. Mehta, Multirate output feedback based discrete-time sliding mode control for fractional delay compensation in NCSs, in *IEEE Conference on Industrial Technology (ICIT-2017)* (2017), pp. 848–853

Chapter 6
Discrete-Time Sliding Mode Controller for NCS Having Random Type Fractional Delay and Single Packet Loss

Abstract In Networked Control System, the behaviour of network delays generally depends on the characteristics of communication medium as well as occupancy of channel by different elements. When large number of sensors, controllers and actuators share their information through the common communication medium, then the network delays and packet losses are random in nature. In this chapter, a novel approach is presented for designing discrete-time sliding mode controller by treating random fractional delay and packet loss separately. The fractional delay that occurs within sampling period while transmitted from sensor to controller and controller to actuator channel is modelled using Poisson's distribution function and is approximated using Thiran's delay approximation technique for designing the discrete-time sliding mode controller. The packet loss that occurs in communication channel between sensor to controller and controller to actuator is treated with Bernoulli's distribution function and compensated at controller end as well as actuator end. Further, Lyapunov approach is used to determine the stability of closed-loop NCSs with proposed discrete-time SMC controller. The feasibility and efficiency of the proposed control methodology is outlined through simulation and experimental results which shows a significant response even in the presence of random fractional delay, packets loss and matched uncertainties.

Keywords Discrete-time sliding mode control · Networked control
Random fractional delay · Packet loss

6.1 Problem Formulation

The block diagram of NCS with communication-networked medium and packet loss situation is shown in Fig. 6.1. The state information as well as control signal are transmitted to the controller and actuator through the network medium. During transmission, the state information and control signal experience time delay from sensor to controller channel and controller to actuator channel, respectively [1, 2]. These delays are broadly defined as the time required for the data packets to travel within the network. While transmission, if data packets take longer duration of time

© Springer Nature Singapore Pte Ltd. 2018
D. H. Shah and A. Mehta, *Discrete-Time Sliding Mode Control for Networked Control System*, Studies in Systems, Decision and Control 132,
https://doi.org/10.1007/978-981-10-7536-0_6

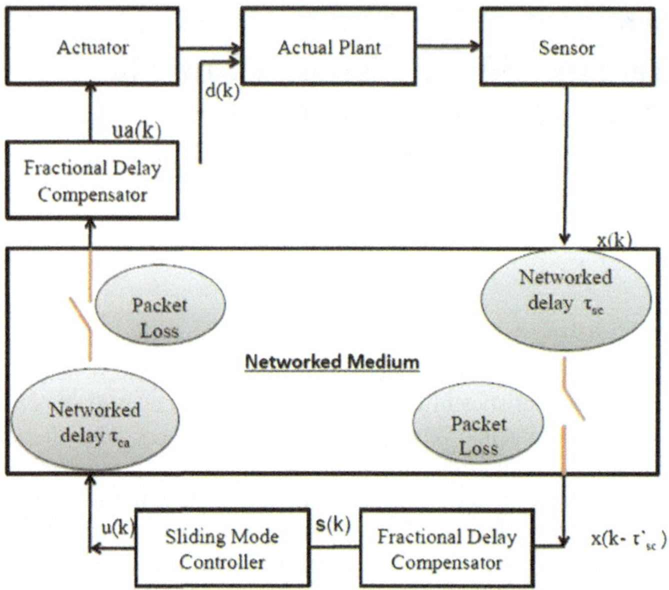

Fig. 6.1 Block diagram of NCS with fractional delay compensation and packet loss

to travel within the network than sampling interval it is called as packet loss [3, 4]. This situation mainly arises due to network congestion, node competition or heavy traffic in the network [5, 6].

Consider the linear time-invariant SISO system with random network delay in continuous-time domain as:

$$\dot{x}(t) = Ax(t) + Bu(t - \tau_r) + Dd(t), \tag{6.1}$$

$$y(t) = Cx(t), \tag{6.2}$$

where $x \in R^n$ is system state vector, $u \in R^m$ is control input, $y \in R^p$ is system output, $A \in R^{n \times n}$, $B \in R^{n \times m}$, $C \in R^{p \times n}$, $D \in R^{n \times m}$ are the matrices of appropriate dimensions, $d(t)$ is the matched bounded disturbance with $|d(t)| \leq d_{max}$, and τ_r is the total random network delay in continuous-time domain.

The discrete form of system (6.1) and (6.2) is:

$$x(k + 1) = Fx(k) + Gu(k - \hat{\tau}) + d(k), \tag{6.3}$$

$$y(k) = Cx(k), \tag{6.4}$$

where $F = e^{Ah}, G = \int_0^h e^{At} B dt, d(k) = \int_0^h e^{At} Dd((k + 1)h - t)dt \in O(h)$. Since $|d(t)| \leq d_{max}$, it can be inferred that $d(k)$ is also bounded and $O(h)$ [7]. For simplicity, it is assumed that $d(k)$ is slowly varying and remains constant over the interval $kh \le t \le (k+1)h$ [7].

The total random fractional delay $(\hat{\tau})$ occurring within the network is denoted as,

$$\hat{\tau} = \frac{\tau_r}{h},$$

where h is the sampling interval.

Remark 15 In this work, the non-integer form of random fractional delay $\hat{\tau}$ is considered instead of integer form in discrete-time domain in order to compensate the precise effect of random network delay occurring at each sampling instants.

Assumption 7 Network-induced fractional delays are random in nature and therefore satisfying

$$\hat{\tau}_l \le \hat{\tau} \le \hat{\tau}_u, \tag{6.5}$$

where $\hat{\tau}_l$ and $\hat{\tau}_u$ indicate the lower and upper bounds of total random fractional network delays.

The total random network-induced fractional delay is the combination of sensor to controller fractional delay $(\hat{\tau}_{sc})$ and controller to actuator fractional delay $(\hat{\tau}_{ca})$ having random behaviour which is given as

$$\hat{\tau} = \hat{\tau}_{sc} + \hat{\tau}_{ca}, \tag{6.6}$$

where $\hat{\tau}_{sc} = \frac{\tau_{rsc}}{h}$ and $\hat{\tau}_{ca} = \frac{\tau_{rca}}{h}$.

Assumption 8 In this work, it is assumed that only single packet loss occurs. The assumption is justifying as the packet time delay more than sampling period is considered as dropped or lost packet.

Assumption 9 The disturbance $d(k)$ is bounded by upper and lower bound as:

$$d_l \le d(k) \le d_u, \tag{6.7}$$

where d_l and d_u denote lower and upper bounds of disturbance.

Problem Statement: To design robust non-switching discrete-time sliding mode controller for the system (6.3, 6.4) in the presence of random fractional delays $\hat{\tau}_{sc}$ and $\hat{\tau}_{ca}$, matched uncertainty and packet loss situation satisfy condition (6.5), (6.7) and Assumption 8.

The sliding mode controller involves the sliding surface design that steers the system states towards the surface and control law that computes the control sequences [8–13].

The next section proposes the design of sliding surface that compensates the effect of random fractional delay occurring between sensor to controller in NCSs. It also discusses the modelling of random fractional delay and packet loss.

6.2 Sliding Surface with Random Fractional Delay and Packet Loss

The effect of random fractional delay in discrete-time domain due to the presence of networked medium is compensated in sliding surface presented in Lemma 4.

Lemma 4 *The sliding surface $s(k)$ for a random fractional delay $(\hat{\tau}_{sc})$, satisfying conditions (6.5) and (6.7) with packet loss for the system (6.3, 6.4), is given as:*

$$s(k) = (1 - \bar{\alpha})x'_c(k) + \bar{\alpha}x'_c(k - 1), \tag{6.8}$$

where
$x'_c(k) = C_s x(k) - \varsigma C_s x(k - 1), x'_c(k - 1) = C_s x(k - 1) - \varsigma C_s x(k - 2), \varsigma = \frac{\hat{\tau}_{sc}}{\hat{\tau}_{sc}+1},$
C_s *is the sliding gain, and $\bar{\alpha}$ is the probability of the data packet lost.*

Proof In discrete-time domain, there are various algorithms for modelling the random variables such as Bernoulli's distribution, geometric distribution, Poisson's distribution, probability distribution, binomial distribution and Pascal distribution. Among these algorithms, Bernoulli's distribution and probability distribution are two widely used algorithms for mathematical modelling of integer type of network delays in discrete-time domain. Since the sensor to controller delay $(\hat{\tau}_{sc})$ is random and fractional in nature, Poisson's distribution is the most suitable approach in discrete-time domain. Poisson's distribution [14] is used to model the random variables of smaller values on the basis of number of events occurred over the specified interval of time. The occurrence of the events is based on the number of trials required to generate the event. So, using this approach the random fractional delay is modelled by assuming that at each sampling instants an event takes place such that the networked delay might be lesser or higher than sampling interval.

Thus, the communicated state variable over the network having random fractional delay from sensor to controller is given as:

$$x_c(k) = x(k - \hat{\tau}_{sc}), \tag{6.9}$$

where $\hat{\tau}_{sc}$ is the fractional form of sensor to controller delay in discrete-time domain that takes the values in a finite set, that is, $\{\hat{\tau}_{sc}\}\varepsilon\{d_1, d_2, \ldots, d_q\}$. Thus, the sensor to controller fractional delay $\{\hat{\tau}_{sc}\}$ can be modelled using Poisson's distribution with probabilities given by,

$$Pr\{\hat{\tau}_{sc} = d_v\} = E\{d_v\} = \beta_v, v = 1, 2, \ldots, q. \tag{6.10}$$

where β_v is the positive scalar and $\Sigma_{v=0}^{q}\beta_v = 1$, $E\{d_v\}$ is the expectation of the stochastic variable d_v. The mathematical representation of β_v with Poisson's distribution is given by:

$$\beta_v = \frac{\lambda^w e^{-\lambda}}{w!}; w = 0, 1, 2, 3, \ldots \tag{6.11}$$

where w indicates the number of trials, λ denotes average number of events per interval, and e denotes the Euler's number.

Applying z-transform to Eq. (6.9), we get,

$$x_c(z) = x(z)z^{-\hat{\tau}_{sc}}, \tag{6.12}$$

where $\hat{\tau}_{sc} = v = \frac{\tau_{rsc}}{h}$.

Using Thiran's approximation [15], $z^{-\hat{\tau}_{sc}}$ can be approximated as

$$z^{-\hat{\tau}_{sc}} = \Sigma_{k=0}^{1}(-1)^k \binom{l}{k}\Pi_{i=0}^{1}\frac{2\hat{\tau}_{sc}+i}{2\hat{\tau}_{sc}+k+i}z^{-k}. \tag{6.13}$$

The above Eq. (6.13) can be further expanded as:

$$z^{-\hat{\tau}_{sc}} = \left[(-1)^0\binom{1}{0}\left\{\frac{2\hat{\tau}_{sc}}{2\hat{\tau}_{sc}}*\frac{2\hat{\tau}_{sc}+1}{2\hat{\tau}_{sc}+1}\right\}z^0 + (-1)^1\binom{1}{1}\right. \tag{6.14}$$
$$\left.\left\{\frac{2\hat{\tau}_{sc}}{2\hat{\tau}_{sc}+1}*\frac{2\hat{\tau}_{sc}+1}{2\hat{\tau}_{sc}+2}\right\}z^{-1}\right].$$

On simplifying, we get

$$z^{-\hat{\tau}_{sc}} = 1 - \varsigma z^{-1}, \tag{6.15}$$

where $\varsigma = \frac{\hat{\tau}_{sc}}{\hat{\tau}_{sc}+1}$ and $\hat{\tau}_{sc}$ is the random fractional delay defined in Eqs. (6.10, 6.11), respectively.

Substituting Eqs. (6.15) into (6.12), we have

$$x_c(z) = x(z)[1 - \varsigma z^{-1}]. \tag{6.16}$$

Further expanding it, we obtain

$$x_c(z) = x(z) - \varsigma z^{-1}x(z). \tag{6.17}$$

Applying inverse z-transform, we get

$$x_c(k) = x(k) - \varsigma x(k-1). \tag{6.18}$$

The sensor output signal $x(k)$ generated at each sampling instant h is sent to the controller via communication channel. From Eq. (6.18), it can be noticed that at each sampling instant the effect of fractional delay ($\hat{\tau}_{sc}$) is compensated in the communicated state variable $x_c(k)$ using the immediate past state, the current state and parameter ς. The same communicated state variable $x_c(k)$ is further used to compute the sliding surface.

The mathematical representation of compensated random packet loss within the network is given as

$$x_p(k) = (1 - \alpha(k))x_c(k) + \alpha(k)x_c(k - 1), \tag{6.19}$$

where $x_c(k)$ is the compensated communicated state variable available at the controller and $\alpha(k) \in R$ is the stochastic variable which is represented as Bernoulli's distributed sequence with

$$Pr\{\alpha(k) = 1\} = E\{\alpha(k)\} = \bar{\alpha}, \tag{6.20}$$

$$Pr\{\alpha(k) = 0\} = 1 - E\{\alpha(k)\} = 1 - \bar{\alpha}, \tag{6.21}$$

where $0 \preceq \bar{\alpha} \prec 1$ implies the probability that the data packet is lost and $E\{\alpha(k)\}$ is the expectation of the stochastic variable $\alpha(k)$.

Thus, $x_p(k)$ can be written as:

$$x_p(k) = (1 - \bar{\alpha})x_c(k) + \bar{\alpha}x_c(k - 1), \tag{6.22}$$

where $\bar{\alpha}$ is the probability of the data packet lost defined in Eqs. (6.20, 6.21).

Let the sliding variable that compensates the effect of random fractional delay and packet loss be:

$$s(k) = C_s x_p(k), \tag{6.23}$$

where C_s is the sliding gain which is calculated using discrete LQR method through proper selection of Q and R matrices.

Substituting the value of $x_p(k)$ in Eq. (6.23), we have:

$$s(k) = (1 - \bar{\alpha})x_c'(k) + \bar{\alpha}x_c'(k - 1), \tag{6.24}$$

where
$x_c'(k) = C_s x(k) - \varsigma C_s x(k - 1),\ x_c'(k - 1) = C_s x(k - 1) - \varsigma C_s x(k - 2)$

This completes the *proof*.

It is assumed that the state packet delay is smaller than sampling interval if the network is free from congestion. So, the data packet $x_p(k)$ with delay is used without loss to compute the sliding surface. However, if the network is overloaded due to traffic or congestion, the state packet delay will be larger than sampling interval. At that instance, $x_p(k - 1)$ will be used to compute the sliding surface. So from Eq. (6.24), it can be easily noticed that at each sampling instant the effect of random fractional delay and packet loss in the actual system states is compensated at the sliding surface when it takes the value $x_c'(k - 1)$ with probability $\bar{\alpha}$ and $x_c'(k)$ with probability $(1 - \bar{\alpha})$.

Once the sliding surface is designed, the next step is to design the discrete-time sliding mode control law which is presented in next section.

6.3 Discrete-Time Networked Sliding Mode Control for NCSs with Random Fractional Delays and Packet Loss

This section presents the derivation of non-switching discrete-time sliding mode control law for NCS using the sliding surface (6.24) as Theorem 6.1 below.

Theorem 6.1 *The non-switching discrete-time sliding mode controller for system (6.3, 6.4) in the presence of random fractional network delays satisfying (6.5) with packet loss and matched uncertainty $d(k)$ is given as*

$$u(k) = -(C_s G)^{-1}(1 - \bar{\alpha})^{-1}[Hx(k) - Ix(k) + Kx(k) - Lx(k-1) \qquad (6.25)$$
$$- J(s(k)) + d_s(k) - d_1] - d(k).$$

where
$H = (1 - \bar{\alpha})(C_s F), I = \varsigma(1 - \bar{\alpha})C_s, K = \bar{\alpha}C_s, L = \varsigma\bar{\alpha}C_s$ and $J = \{1 - q[s(k)]\}$.

Proof Let us define reaching law given by [16] in the presence of random fractional delay as:

$$s[(k+1)h] = \{1 - q[s(k)]\} - d_s(k) + d_1, \qquad (6.26)$$

where $q[s(k)] = \frac{\psi}{\psi+|s(k)|}$ with ψ as user-defined constant satisfying $\psi \geq d_2$, and d_1 and d_2 are mean and deviation of $d(k)$.

Remark 16 The disturbance $d(k)$ appearing in the reaching law is applied through the network. So, the compensated disturbance $d_s(k)$ using Thiran's approximation [15] is given as:

$$d_s(k) = d(k) - \varsigma d(k-1). \qquad (6.27)$$

The reaching law in Eq. (6.26) indicates that the system states always move towards the specified sliding band given as:

$$|s(k)| \leq \frac{\psi d_2}{\psi - d_2}, \qquad (6.28)$$

Substituting the value of $s(k+1)$ in Eq. (6.26), we get

$$(1 - \bar{\alpha})x_c'(k+1) + \bar{\alpha}x_c'(k) = \{1 - q[s(k)]\} - d_s(k) + d_1,$$

Substituting the value of $x_c'(k+1)$,

$$(1 - \bar{\alpha})[C_s x(k+1) - \varsigma C_s x(k)] + \bar{\alpha}[C_s x(k) - \varsigma C_s x(k-1)] \qquad (6.29)$$
$$= \{1 - q[s(k)]\} - d_s(k) + d_1.$$

Remark 17 It is noticed from Eq. (6.24) that random sensor to controller fractional delay is compensated at the sliding surface, while random controller to actuator fractional delay is compensated at the actuator side. So, the effect of delay will not be observed at the controller side for computing the control signal as well at actuator side. Thus, without loss of generality, the control signal in Eq. (6.3) is given as

$$u(k - \tau') = u(k). \tag{6.30}$$

Further, substituting $x(k + 1)$ gives

$$(1 - \bar{\alpha})[C_s[Fx(k) + G(u(k) + d(k))] - \varsigma C_s x(k)] + \bar{\alpha}[C_s x(k) \tag{6.31}$$
$$- \varsigma C_s x(k - 1)] = \{1 - q[s(k)]\} - d_s(k) + d_1.$$

Simplifying,

$$(1 - \bar{\alpha})C_s Fx(k) + (1 - \bar{\alpha})C_s G(u(k) + d(k)) - \varsigma(1 - \bar{\alpha})C_s x(k) \tag{6.32}$$
$$+ \bar{\alpha}C_s x(k) - \varsigma \bar{\alpha}C_s x(k - 1) = \{1 - q[s(k)]\} - d_s(k) + d_1.$$

Further solving above Eq. (6.32), the control law can be expressed as:

$$u(k) = -(C_s G)^{-1}(1 - \bar{\alpha})^{-1}[Hx(k) - Ix(k) + Kx(k) - Lx(k - 1) \tag{6.33}$$
$$- J(s(k)) + d_s(k) - d_1] - d(k),$$

where $H = (1 - \bar{\alpha})(C_s F)$, $I = \varsigma(1 - \bar{\alpha})C_s$, $K = \bar{\alpha}C_s$, $L = \varsigma \bar{\alpha}C_s$ and $J = \{1 - q[s(k)]\}$.

This completes the *proof*.

Similarly, the effect of random fractional delay at controller to actuator and packet loss can be compensated at the actuator end. The compensated control signal applied to the plant is given by:

$$u_a(k) = (1 - \bar{\beta})(u_c(k)) + \bar{\beta}(u_c(k - 1)), \tag{6.34}$$

where
$u_c(k) = u(k) - \gamma'u(k - 1)$ and $u_c(k - 1) = u(k - 1) - \gamma'u(k - 2)$ and $\gamma' = \frac{\hat{t}_{ca}}{1 + \hat{t}_{ca}}$,
and $\hat{t}_{ca}$ is the random fractional delay from controller to actuator.

From Eq. (6.34), it can be easily inferred that at each sampling instant the effect of random fractional delay from controller to actuator is compensated in $u_c(k)$ using past control signal, present control signal and parameter "γ'", while packet loss is compensated which takes the value $u_c(k - 1)$ with probability $\bar{\beta}$ and $u_c(k)$ with probability $(1 - \bar{\beta})$.

The next section discusses the stability condition for the closed-loop system such that the system states remain within specified band (6.28) using control law (6.33).

6.4 Stability Analysis

Theorem 6.2 *The trajectories of the closed-loop system given in Eqs. (6.3, 6.4) drive towards the sliding surface as mentioned in Eq. (6.24) for a given controller in Eq. (6.33) in the presence of total networked delay $\hat{\tau}$ satisfying Eq. (6.5), packet loss condition satisfying $0 \preceq \bar{\alpha} \prec 1$ and matched uncertainty satisfying (6.7) such that the following condition in Eq. (6.35) must exist:*

$$\rho s^T(k)s(k) \succ 0. \tag{6.35}$$

Proof Consider the compensated sliding surface (6.24)

$$s(k) = (1 - \bar{\alpha})x_c'(k) + \bar{\alpha}x_c'(k - 1). \tag{6.36}$$

Let us define Lyapunov function as

$$V_s(k) = s^T(k)s(k). \tag{6.37}$$

Taking the forward difference, we have

$$\Delta V_s(k) = s^T(k + 1)s(k + 1) - s^T(k)s(k). \tag{6.38}$$

Using Eq. (6.36), we get

$$\Delta V_s(k) = [(1 - \bar{\alpha})x_c'(k + 1) + \bar{\alpha}x_c'(k)]^T[(1 - \bar{\alpha})x_c'(k + 1) + \bar{\alpha}x_c'(k)] - s^T(k)s(k). \tag{6.39}$$

Substituting the value of $x_c'(k + 1)$, we get

$$\Delta V_s(k) = [(1 - \bar{\alpha})[C_s x(k + 1) - \alpha'C_s x(k)] + \bar{\alpha}[C_s x(k) - \alpha'C_s x(k - 1)]]^T \\ [(1 - \bar{\alpha})[C_s x(k + 1) - \alpha'C_s x(k)] + \bar{\alpha}[C_s x(k) - \alpha'C_s x(k - 1)]] - s^T(k)s(k). \tag{6.40}$$

Further substituting the value of $x(k + 1)$,

$$\Delta V_s(k) = [(1 - \bar{\alpha})[C_s[Fx(k) + G(u(k) + d(k))] - \alpha'C_s x(k)] + \bar{\alpha}[C_s x(k) \\ - \alpha'C_s x(k - 1)]]^T[(1 - \bar{\alpha})[C_s[Fx(k) + G(u(k) + d(k))] - \alpha'C_s x(k)] \\ + \bar{\alpha}[C_s x(k) - \alpha'C_s x(k - 1)]] - s^T(k)s(k). \tag{6.41}$$

Substituting $u(k)$ and rewriting above Eq. (6.39),

$$\Delta V_s(k) = \Gamma - s^T(k)s(k), \tag{6.42}$$

where $\quad \Gamma = [(1 - \bar{\alpha})^{-1}[1 - q(s(k))]s(k) - d_s(k) + d_1]^T[(1 - \bar{\alpha})^{-1}[1 - q(s(k))]s(k) - d_s(k) + d_1]$. The term Γ can be tuned close to zero by appropriately selecting the parameter ψ and $\bar{\alpha}$. If Γ is close to zero, then $s^T(k)s(k)$ will be larger than Γ. Thus, for any small parameter ρ we have

$$\Gamma - s^T(k)s(k) \prec \rho s^T(k)s(k). \tag{6.43}$$

Tuning of parameter Γ leads to $\Delta V_s(k) \prec \rho s^T(k)s(k)$ which guarantees the convergence of $\Delta V_s(k)$ and implies that any trajectory of the system (6.3, 6.4) will be driven onto the sliding surface and maintain on it.

This completes the *proof*.

6.5 Results and Discussions

In this section, the effect of proposed control algorithm is validated through simulation and experimental results carried out in the presence of random fractional delay, packet loss and matched uncertainty. An illustrative example given by [17] is considered for simulation, while Quanser DC motor is used for experimental purpose.

6.5.1 Simulation Results

Consider the continuous-time LTI system as

$$\dot{x}(t) = Ax(t) + Bu(t - \tau_r) + Dd(t), \tag{6.44}$$
$$y(t) = Cx(t), \tag{6.45}$$

where
$$A = \begin{bmatrix} -0.7 & 2 \\ 0 & -1.5 \end{bmatrix}, B = \begin{bmatrix} -0.03 \\ -1 \end{bmatrix},$$

$$C = \begin{bmatrix} 1 & 0 \end{bmatrix}, D = \begin{bmatrix} 1 \\ 1 \end{bmatrix}, d(t) = 0.2\sin(0.086t).$$

Discretizing the above system with sampling interval of $h = 30\,\text{ms}$, we get

$$x(k + 1) = Fx(k) + Gu(k - \hat{\tau}) + d(k), \tag{6.46}$$
$$y(k) = Cx(k), \tag{6.47}$$

where
$$F = \begin{bmatrix} 0.9792 & 0.05805 \\ 0 & 0.956 \end{bmatrix}, G = \begin{bmatrix} -0.001771 \\ -0.02934 \end{bmatrix},$$

$$C = \begin{bmatrix} 1 & 0 \end{bmatrix}.$$

Figures 6.2, 6.3, 6.4, 6.5, 6.6, 6.7, 6.8, 6.9, 6.10, 6.11, 6.12, 6.13, 6.14, 6.15, 6.16, 6.17, 6.18, 6.19, 6.20, 6.21 and 6.22 show the nature of the system under networked environment with random fractional delays, packet loss situation and matched uncertainty. In order to prove the robustness of the proposed algorithm in the presence of variable networked delay and packet loss, the slow time-varying disturbance is applied to the system. The random nature of total networked-induced fractional delay is modelled using Poisson's distribution. It is assumed that at every sampling instant one event is generated considering total networked-induced delay lesser than sampling interval with zero trial. So, according to Poisson's distribution under these assumptions the probability of networked delays lesser than sampling interval will be $p = 0.63$, while the probability of the networked delays greater than sampling interval is $1 - p = 0.37$. Thus, the total network-induced fractional delay generated within the network is $0.003\,\mathrm{s} \leq \tau' \leq 0.055\,\mathrm{s}$, respectively. Figure 6.2 depicts the random nature of total networked-induced fractional delays modelled using Poisson's distribution. Figures 6.3 and 6.4 show the magnified results of sensor to controller fractional delay and controller to actuator fractional delay. As discussed in previous chapter, the processing delays and computational delays occurring at sensor, actuator and controller are neglected due to its negligible effect.

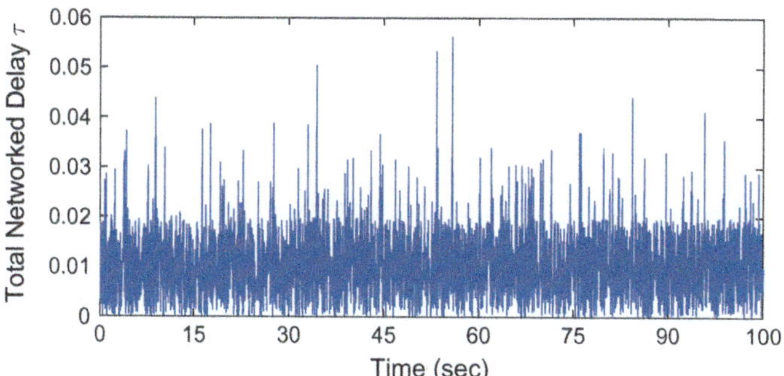

Fig. 6.2 Total networked fractional delay τ_r

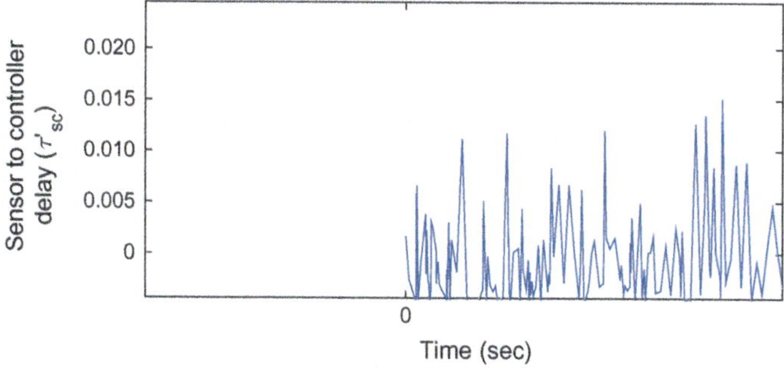

Fig. 6.3 Magnified sensor to controller fractional delay

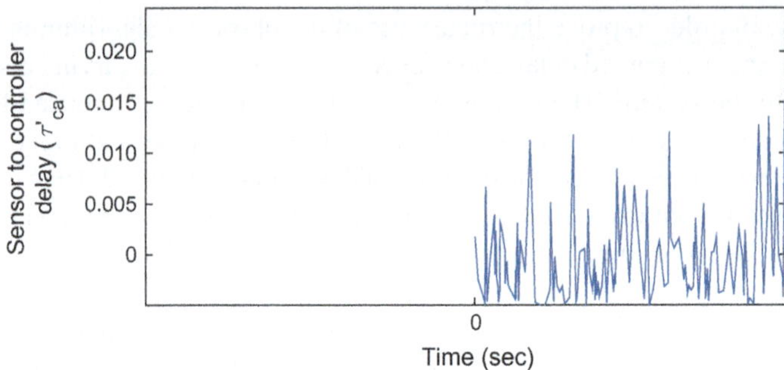

Fig. 6.4 Magnified controller to actuator fractional delay

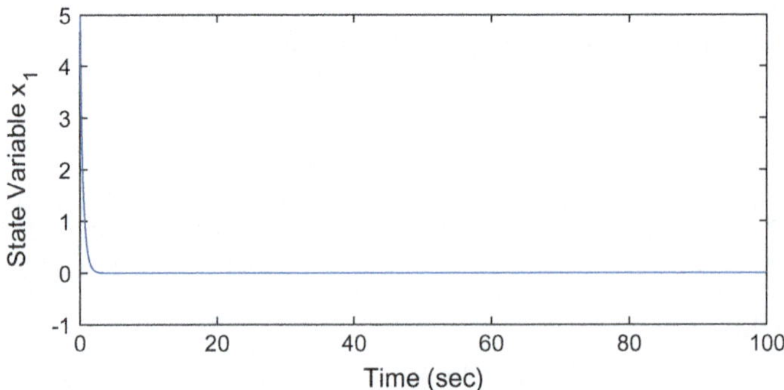

Fig. 6.5 State variable $x_1(k)$

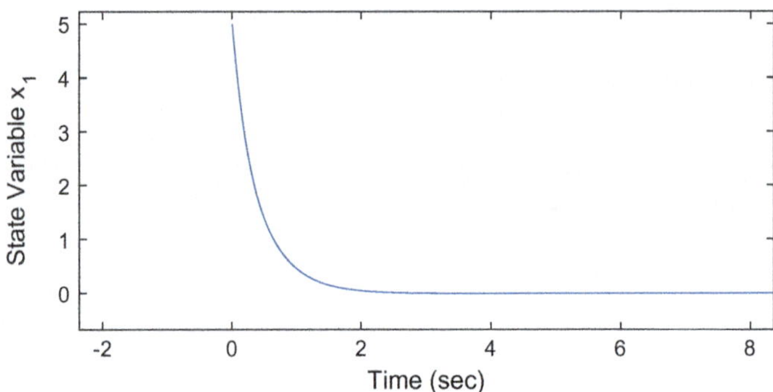

Fig. 6.6 Magnified state variable $x_1(k)$

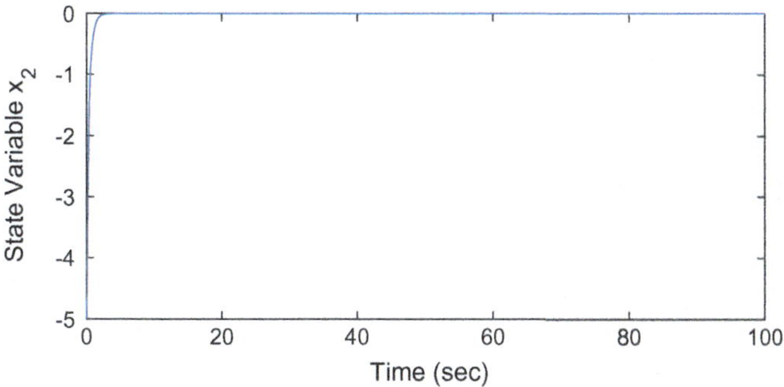

Fig. 6.7 State variable $x_2(k)$

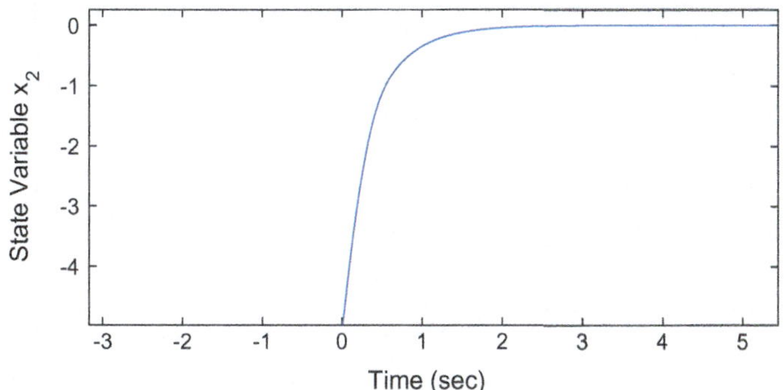

Fig. 6.8 Magnified state variable $x_2(k)$

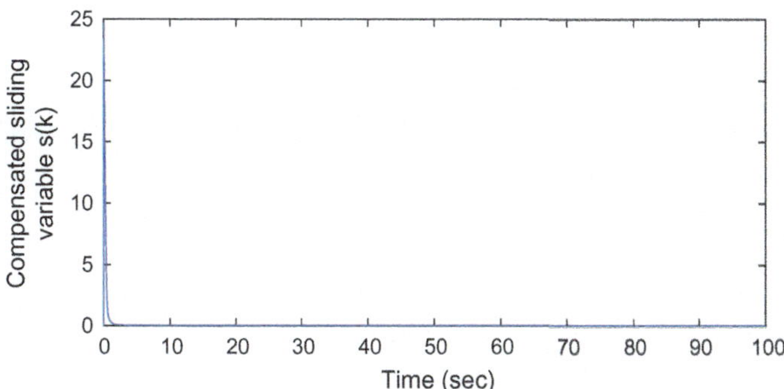

Fig. 6.9 Compensated sliding variable $s(k)$

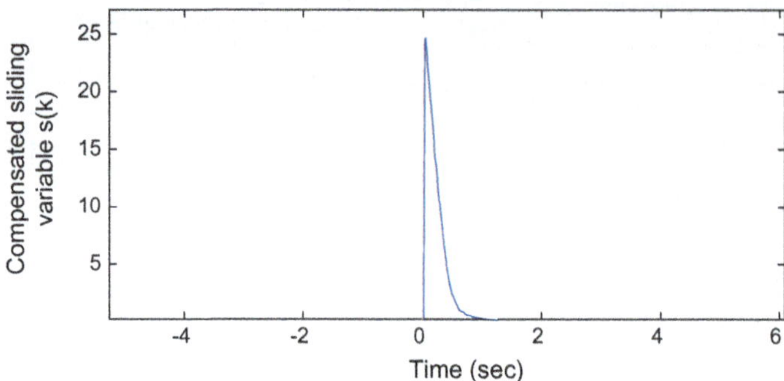

Fig. 6.10 Magnified compensated sliding variable $s(k)$

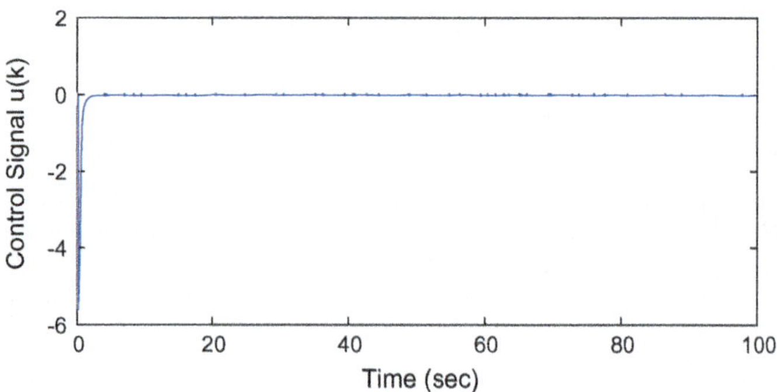

Fig. 6.11 Control signal $u(k)$

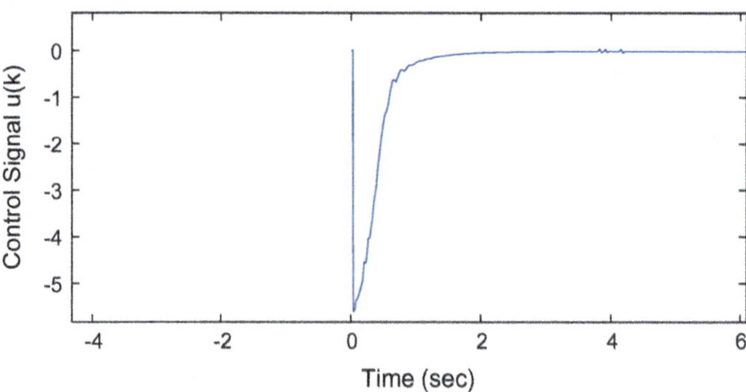

Fig. 6.12 Magnified control signal $u(k)$

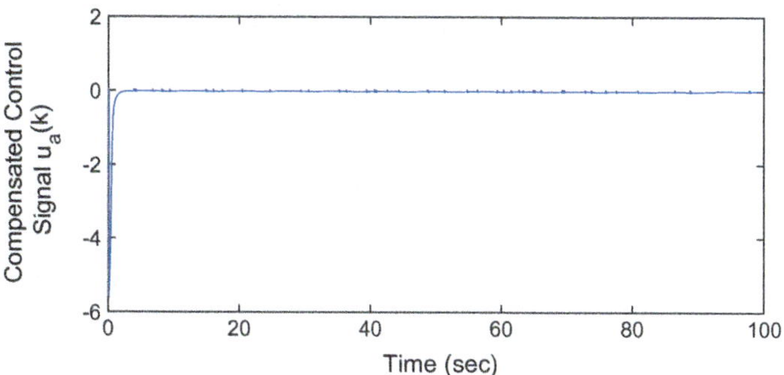

Fig. 6.13 Compensated control signal $u_a(k)$

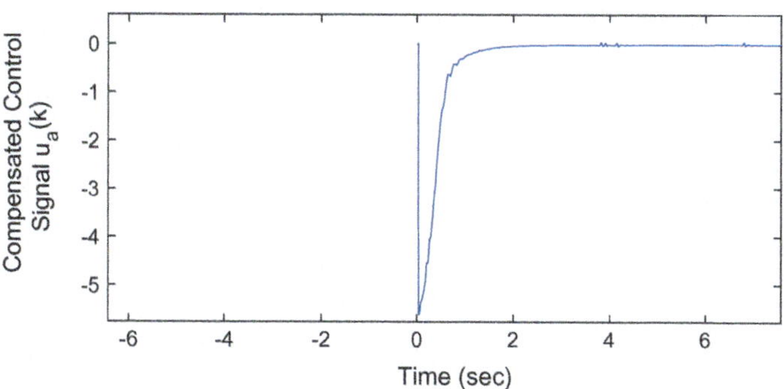

Fig. 6.14 Magnified compensated control signal $u_a(k)$

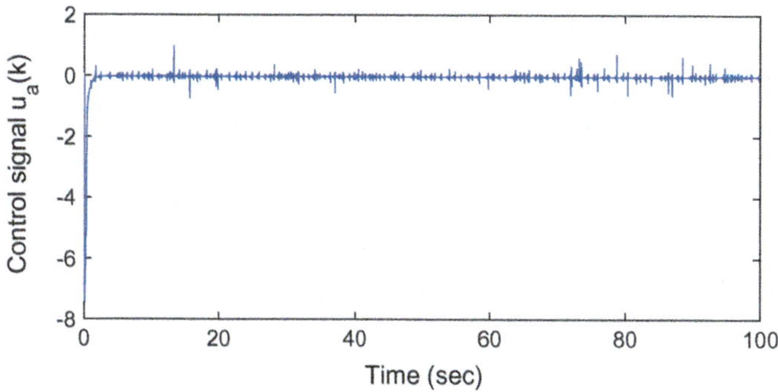

Fig. 6.15 Compensated control signal $u_a(k)$ with 10% packet loss

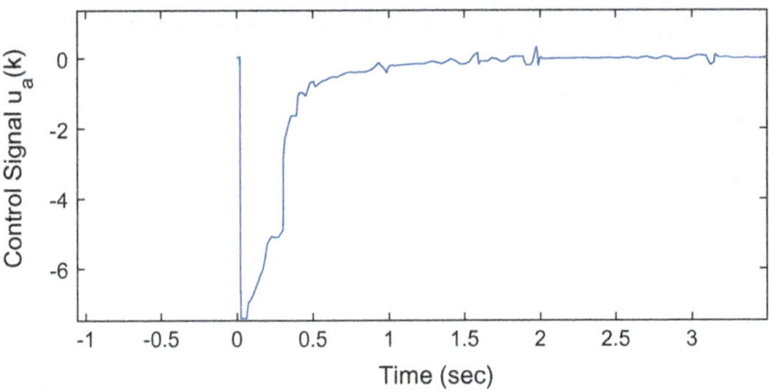

Fig. 6.16 Magnified compensated control signal $u_a(k)$ with 10% packet loss

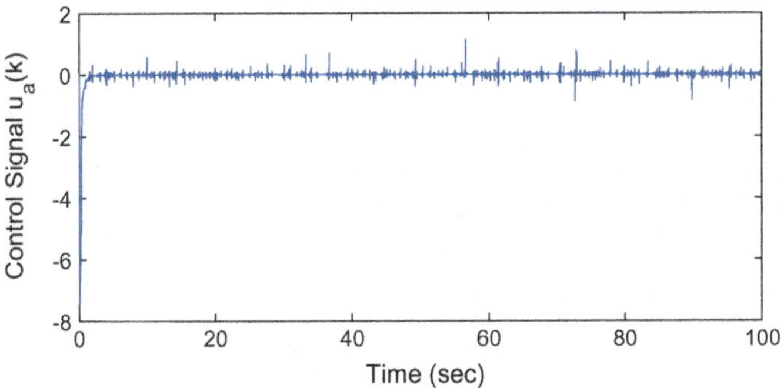

Fig. 6.17 Compensated control signal $u_a(k)$ with 30% packet loss

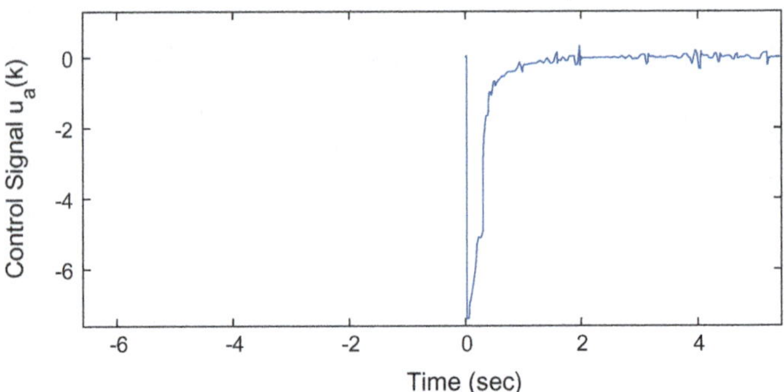

Fig. 6.18 Magnified compensated control signal $u_a(k)$ with 30% packet loss

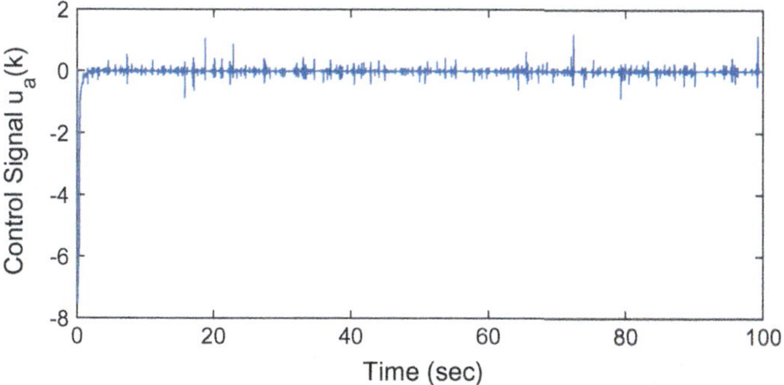

Fig. 6.19 Compensated control signal $u_a(k)$ with 50% packet loss

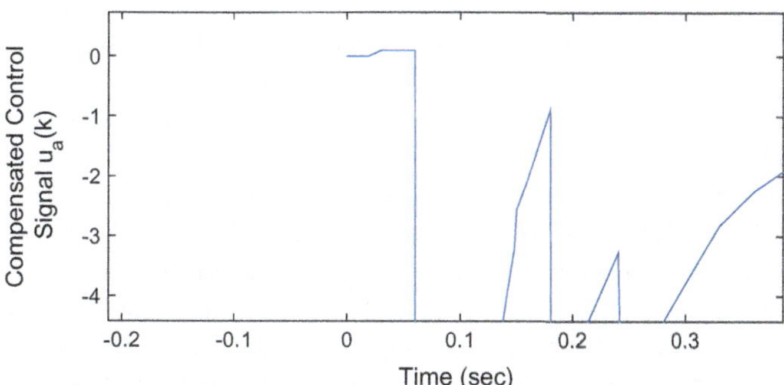

Fig. 6.20 Magnified compensated control signal $u_a(k)$ with 50% packet loss

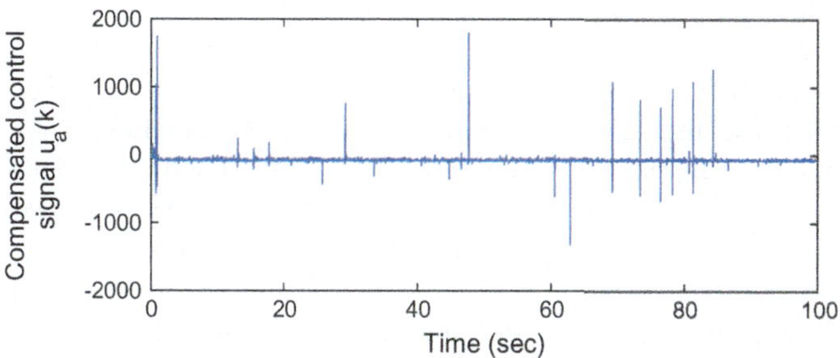

Fig. 6.21 Compensated control signal $u_a(k)$ with fractional delay greater than sampling interval

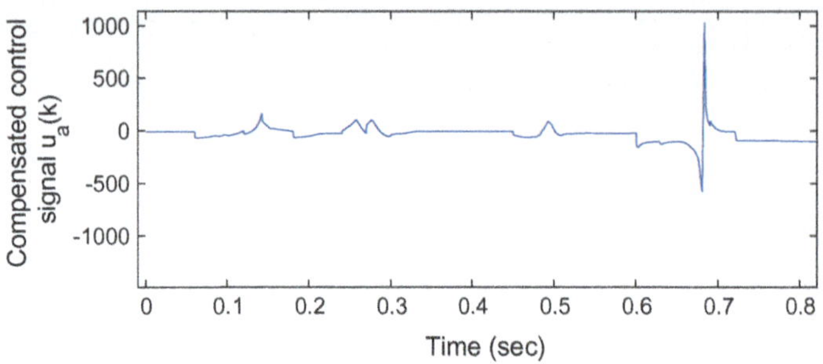

Fig. 6.22 Magnified compensated control signal $u_a(k)$ with fractional delay greater than sampling interval

The sliding gain C_s is calculated using discrete LQR method with $Q = diag(1500, 1000)$ and $R = 1$. The computed values of sliding gain come out to be $C_s = [-1.577 \ -2.456]$. The quasi-sliding mode band comes out to be $|s(k)| \preceq +0.1$ to -0.1 with proper selection of user-defined constant $\psi = 10$.

The simulation results are discussed in three parts: (i) Figs. 6.5, 6.6, 6.7, 6.8, 6.9, 6.10, 6.11, 6.12, 6.13 and 6.14 discuss the effect of proposed control algorithm for specified networked delay range under single packet transmission, (ii) Figs. 6.15, 6.16, 6.17, 6.18, 6.19 and 6.20 describe the effect of fractional delays on compensated control signal for different packet loss situations, and (iii) Figs. 6.21 and 6.22 show the results of compensated control signal computed at actuator side when fractional delays are greater than the sampling interval.

Figures 6.5 and 6.7 show the results of system state variables with initial condition $x(k) = [5 \ -5]$. It can be noticed that both the state variables slide towards the origin from given initial condition in the presence of random fractional delays. In order to show the accurate effect of random fractional delay compensation at the output, results are magnified as shown in Figs. 6.6 and 6.8, respectively. It can be observed that in both the cases the state variables are computed from initial sampling. Thus, the effect of random fractional delay from sensor to controller and controller to actuator is compensated using the proposed algorithm. The same effect of compensation is observed in compensated sliding variable (Fig. 6.9), control signal (Fig. 6.11) and compensated control signal (Fig. 6.13), respectively. On observation of the magnified results in Figs. 6.10, 6.12 and 6.14, it can be noticed that all the three parameters are computed from first sampling instant even in the presence of sensor to controller delay and controller to actuator delay. The amount of delays generated at both sides of network at that sampling instant is indicated in Figs. 6.3 and 6.4, respectively. Thus, the effects of random fractional delay from sensor to controller are compensated in the sliding surface and remain within the specified band (6.28), while controller to actuator random fractional delay is compensated at the actuator end. The efficacy of the proposed algorithm was further tested under packet loss situation for different probabilities. Figures 6.15, 6.17 and 6.19 show the results of compensated control signal for different values of $\bar{\alpha} = 0.10, 0.30$ and 0.50, respectively. It can be noticed that when the packet loss probability within the network is 50% the system shows unacceptable response with high-frequency oscillations. Thus, it can be noticed that when the packet loss is generated within the network the robust terms will generate more action to stabilize the system which in turn makes the compensated control signal oscillatory in nature. Figures 6.16, 6.18 and 6.20 show the magnified results of the compensated control signal for a given set of packet loss probabilities. It can be observed that when the packet loss probability increases to 50% within the network the proposed algorithm cannot compensate the effect of fractional delay as compared to the cases of 10 and 30% packet losses. Figures 6.21 and 6.22 show the nature of compensated control signal when the probabilities of the random fractional delays are greater than the sampling interval. It can be observed that when the probabilities of network fractional delays are reversed than previous case that is $p = 0.37$ and $1 - p = 0.63$ the proposed technique cannot compensate the effect of random fractional delays and the system response becomes unstable with high-frequency oscillations generated at the output. Thus, from above results it can be concluded that the proposed technique works efficiently with random networked delay of $0.003\,\mathrm{s} \le \hat{\tau} \le 0.055\,\mathrm{s}$ in simulated environment. The proposed controller compensates the effect of random networked delay for $\psi = 10$ and $\bar{\alpha} = 0.3$ satisfying Eq. (6.5) and shows the stable response satisfying condition mentioned in Eq. (6.35) in the presence of random fractional delays, packet loss and matched uncertainty.

6.5.2 *Experimental Results*

The state space form of DC motor plant [18] is given as

$$\dot{x}(t) = Ax(t) + Bu(t - \tau_r) + Dd(t), \tag{6.48}$$
$$y(t) = Cx(t), \tag{6.49}$$

where
$$A = \begin{bmatrix} -201 & 0 \\ 1 & 0 \end{bmatrix}, B = \begin{bmatrix} 1 \\ 0 \end{bmatrix},$$

$$C = \begin{bmatrix} 0 & 1 \end{bmatrix}, D = \begin{bmatrix} 1 \\ 1 \end{bmatrix}, d(t) = 0.2\sin(0.086t).$$

Discretizing the above system with sampling interval of $h = 30\,\mathrm{ms}$, we get

$$x(k + 1) = Fx(k) + Gu(k - \hat{\tau}) + d(k), \tag{6.50}$$
$$y(k) = Cx(k), \tag{6.51}$$

where

$$F = \begin{bmatrix} 0.001836 & 0 \\ 0.004573 & 1 \end{bmatrix}, G = \begin{bmatrix} -0.004753 \\ -0.0001242 \end{bmatrix},$$

$$C = \begin{bmatrix} 0 & 1 \end{bmatrix}.$$

In this section, experimental results are briefly discussed with DC motor as a plant in the presence of random fractional delays, packet loss and matched uncertainty situations. The position of DC motor is considered as a reference signal. The variable fractional delays are computed using Poisson's distribution considering the same assumptions as mentioned in the simulation results section. The values of sliding gain parameter C_s, sliding band $|s(k)|$ and user-defined constant ψ are same in order to study the effect of proposed control algorithm on the real-time system. In simulation section, the effects of control algorithm are well explained under two different conditions (i) in the absence of packet loss and (ii) in the presence of packet loss. But, when any system is connected to real-time networks there are very few chances of secure communication between the plant and controller in terms of data transfer. Some of the data packets will be lost due to various reasons such as jitter, congestion or traffic problems within the network. So, in real-time application it is essential to study the effect of control algorithm in the presence of packet loss and random time delay. Figures 6.23, 6.24, 6.25, 6.26, 6.27, 6.28, 6.29, 6.30, 6.31, 6.32, 6.33 and 6.34 show the nature of the DC motor plant in terms of reference tracking, compensated sliding variable, control signal and compensated control signal for specified random fractional delays shown in Fig. 6.2 and matched uncertainty shown in Fig. 4.2 with probability of single packet loss $\bar{\alpha} = 0.1$. Figure 6.23 shows the tracking response of the DC motor. It can be observed that the position of DC motor is controlled accurately in the presence of random fractional delays and packet loss situation. In order to show the effect of time delay compensation, the magnified tracking result is shown in Fig. 6.24. It can be noticed that the effect of total network fractional delay is compensated as the output tracks the reference signal at first sampling instant. The same effect of fractional delay compensation is observed in compensated sliding variable and control signal as shown in Figs. 6.25 and 6.27, respectively. Observing the magnified results shown in Figs. 6.26 and 6.28, it can be noticed that both the parameters are computed from first sampling instant. Thus, the effect of random fractional delay from sensor to controller is compensated at the sliding surface. The magnified result of random fractional delay from sensor to controller is shown in Fig. 6.3. Figure 6.29 shows the nature of compensated control signal. It can be noticed that the effect of random fractional delay from controller to actuator is compensated at the actuator side and converges to zero rapidly without increasing the amplitude. The magnified result of the same is shown in Fig. 6.30. The magnified result of controller to actuator delay is shown in Fig. 6.4. The efficacy of the proposed algorithm was further tested by increasing the packet loss probability by three times, that is, $\bar{\alpha} = 0.3$ for specified network delays and matched uncertainty. Observing the results of compensated control signal as shown

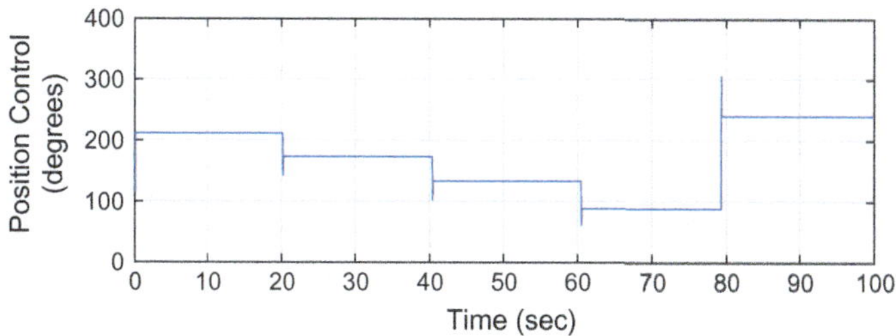

Fig. 6.23 Results of position control of DC motor with 10% packet loss

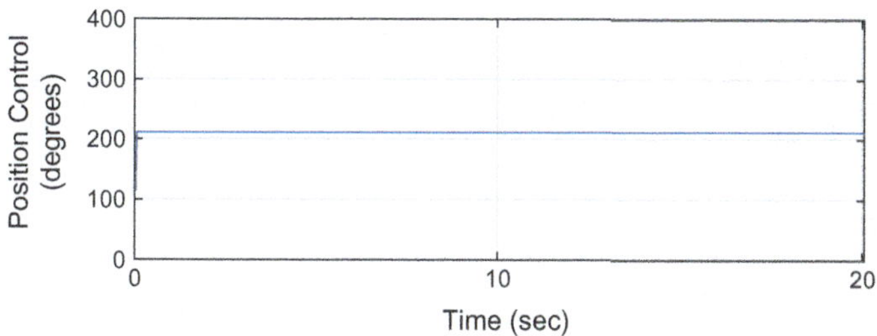

Fig. 6.24 Result of magnified position control of DC motor with 10% packet loss

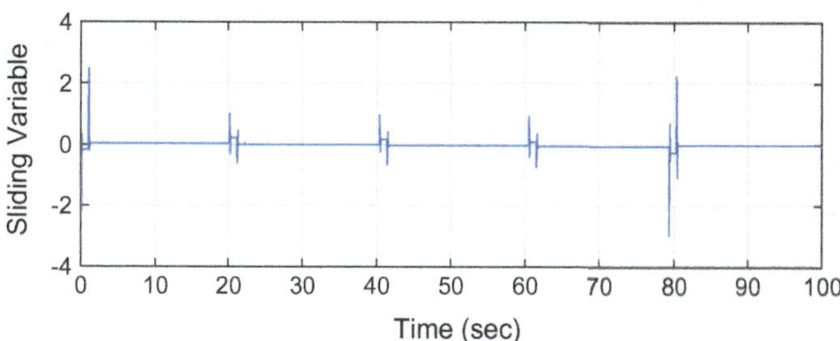

Fig. 6.25 Result of compensated sliding variable with 10% packet loss

in Figs. 6.31 and 6.32, it can be noticed that the effect of fractional delay at the actuator side is still compensated with packet loss probability of 30%. Figure 6.33 shows the nature of tracking response when random fractional delay is greater than sampling interval. It can be noticed that the performance of the system goes to unstable condition and does not compensate the effect of random fractional delays. The magnified result of the same is shown in Fig. 6.34.

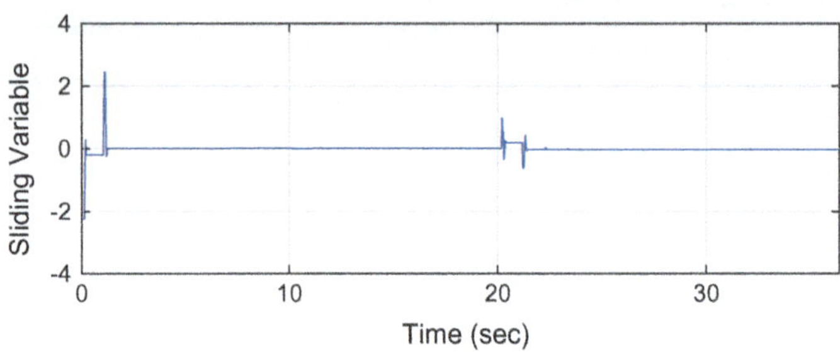

Fig. 6.26 Magnified compensated sliding variable with 10% packet loss

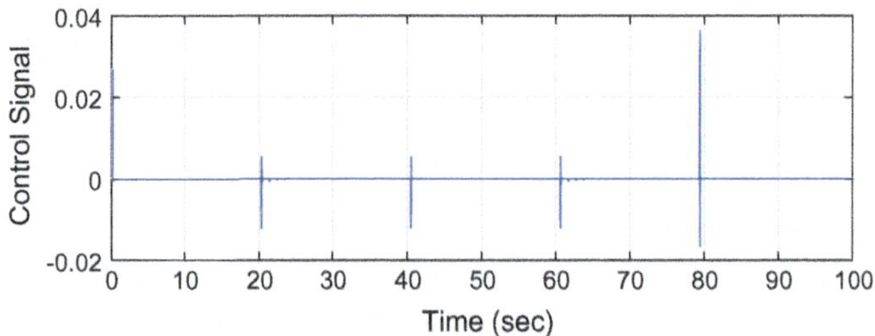

Fig. 6.27 Control signal with 10% packet loss

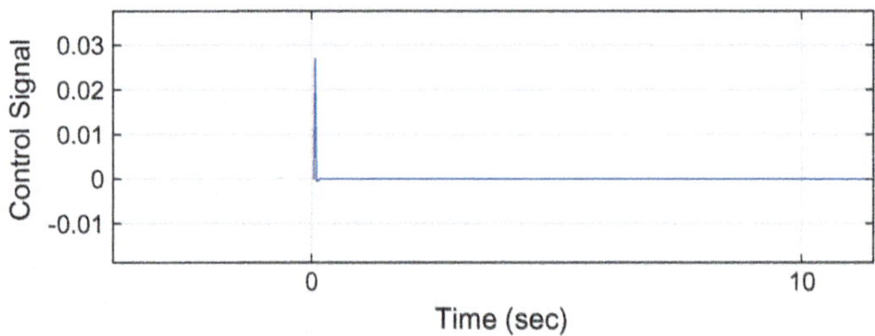

Fig. 6.28 Magnified control signal with 10% packet loss

Thus from above implementation results, it can be noticed that the proposed control algorithm proves to be robust and efficient controller as it shows the stable response satisfying condition (6.35) and compensates the effect of specified random fractional delays satisfying (6.5) in the presence of packet loss and matched uncertainty.

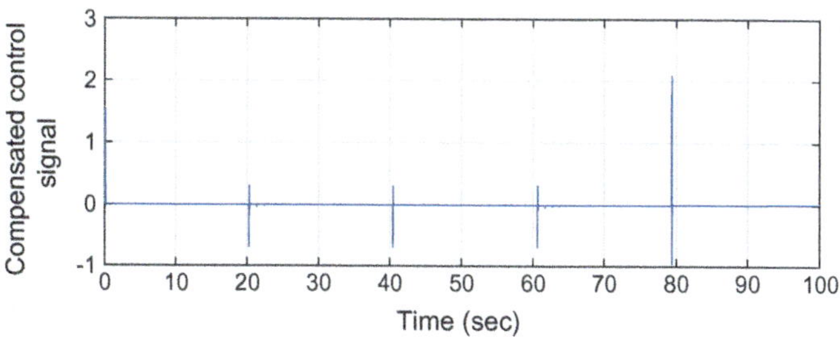

Fig. 6.29 Compensated control signal $u_a(k)$ with 10% packet loss

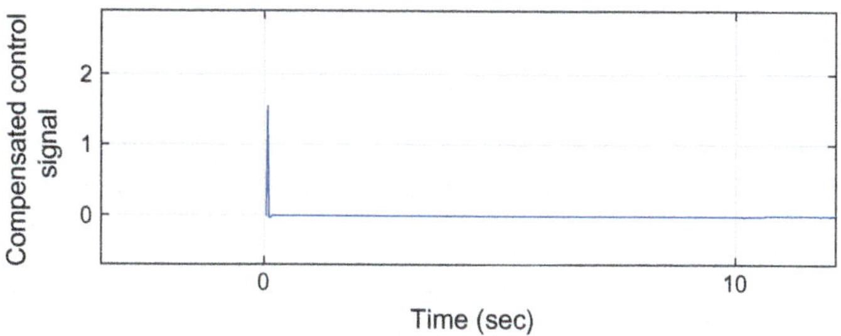

Fig. 6.30 Magnified compensated control signal $u_a(k)$ with 10% packet loss

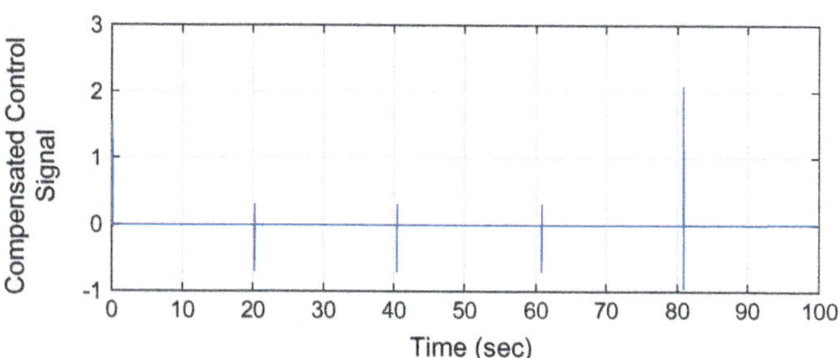

Fig. 6.31 Compensated control signal $u_a(k)$ with 30% packet loss

6.6 Conclusion

In this chapter, we proposed the discrete-time SMC (non-switching) algorithm in the presence of random fractional delay and packet loss in the presence of uncertainty. The random fractional delay is modelled using Poisson's distribution, and packet loss is modelled using Bernoulli's distribution function. The random fractional delay in forward and feedback channels is compensated by Thiran's approximation in the

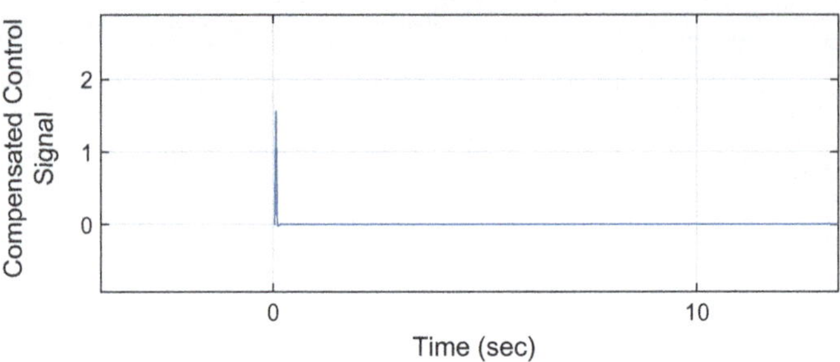

Fig. 6.32 Magnified compensated control signal $u_a(k)$ with 30% packet loss

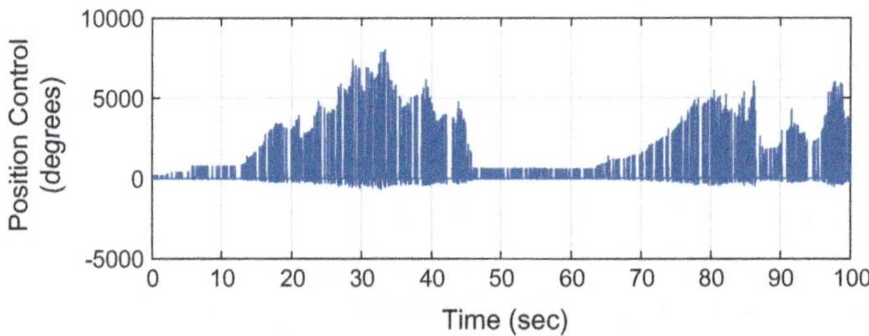

Fig. 6.33 Tracking response with network delays greater than sampling interval

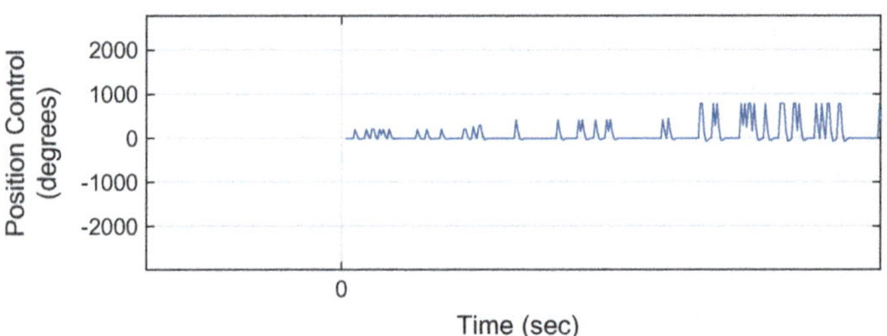

Fig. 6.34 Magnified tracking response with network delays greater than sampling interval

sliding surface and actuator end, respectively. A novel sliding surface is designed using Thiran's approximation. A non-switching type discrete-time sliding mode controller is designed such that system states slide along the proposed compensated surface and maintain within the specified band. The stability condition of closed-loop NCSs is derived using Lyapunov approach that ensures finite-time convergence of system states in the presence of network non-idealities. The effectiveness of the proposed algorithm is examined under different possible conditions through illustrative example as well as real-time plant. The results proved that the control law derived using Thiran's Approximation compensates the random fractional delay precisely even in the presence of 30% packet loss as well as networked delay having values greater than sampling interval.

References

1. J. Nilsson, B. Bernhardsson, B. Wittenmark, Stochastic analysis and control of real time systems with random time delays. Elsevier **2**(3), 13–20 (1997)
2. L. Zhang, Y. Shi, T. Chen, B. Huang, A new method for stabilization of networked control systems with random delays. IEEE Trans. Autom. Control **50**(8), 1177–1181 (2005)
3. H. Song, S.-C. Chen, Y. Yam Sliding mode control for discrete-time systems with Markovian packet dropouts. IEEE Trans. Cybern. **99**, 1–11 (2016)
4. M.A. Khanesar, O. Kaynak, S. Yin, H. Gao, Adaptive indirect fuzzy sliding mode controller for networked control systems subject to time varying network induced time delay. IEEE Trans. Fuzzy Syst. **23**(1), 1–10 (2014)
5. Y. Niu, D.W.C. Ho, Design of sliding mode control subject to packet losses. IEEE Trans. Autom. Control **55**(11), 2623–2628 (2010)
6. A. Argha, L. Li, W. Su Steven, H. Nguyen, Discrete-time sliding mode control for networked systems with random communication delays, in *Proceedings of American Control Conference* (2015), pp. 6016–6021
7. A.J Mehta, B. Bandyopadhyay, Multirate output feedback based stochastic sliding mode control. J. Dyn. Syst. Measur. Control **138**(12), 124503(1–6) (2016)
8. D. Shah, A. Mehta, Multirate output feedback based discrete-time sliding mode control for fractional delay compensation in NCSs, in *IEEE Conference on Industrial Technology (ICIT-2017)* (2017), pp. 848–853
9. D. Shah, A. Mehta, Discrete-time sliding mode control using Thiran's delay approximation for networked control system, in *43rd Annual Conference on Industrial Electronics (IECON-17)*, pp. 3025–3031, Nov. 2017
10. D. Shah, A.J Mehta, Design of robust controller for networked control system, in *Proceedings of IEEE International Conference on Computer, Communication and Control Technology* (2014), pp. 385–390
11. D. Shah, A. Mehta, Output feedback discrete-time networked sliding mode control, in *IEEE Proceedings of Recent Advances in Sliding Modes (RASM)* (2015), pp. 1–7
12. D. Shah, A. Mehta, Discrete-time sliding mode controller subject to real-time fractional delays and packet losses for networked control system. Int. J. Control Autom. Sys. (IJCAS) **15**(6), 2690–2703 (Dec. 2017)
13. D. Shah, A. Mehta, Fractional delay compensated discrete-time SMC for networked control system. In *Digital Communication Networks (DCN)*, vol. 2, No. 3 (Elsevier, 2016), pp. 385–390
14. L. Brown, L. Zhao, A test of Poisson's distribution. J. Stat. **64**(3), 611–625 (2002)
15. J. Thiran, Recursive digital filters with maximally flat group delay. IEEE Trans. Circuit Theory **18**(6), 659–664 (1971)
16. A. Bartoszewicz, P. Lesniewski, Reaching law approach to the sliding mode control of periodic review inventory systems. IEEE Trans. Autom. Sci. Eng. **11**(3), 810–817 (2014)
17. J. Wu, T. Chen, Design of networked control systems with packet deopouts. IEEE Trans. Autom. Control **52**(7), 1314–1319 (2007)
18. K. Astrom, J. Apkarian, P. Karam, M. Levis, J. Falcon, *Student Workbook: QNET DC Motor Control Trainer for NI ELVIS* (Quanser, 2015)

Chapter 7
Discrete-Time Sliding Mode Control with Disturbance Estimator for NCS Having Random Fractional Delay and Multiple Packet Loss

Abstract The concept of remotely controlling a system through communication network gave birth to Networked Control Systems (NCSs). The NCSs are traditional feedback control loops closed through a real-time communication network. It is evident that the performance of closed-loop system deteriorates due to network delay and information loss in communication channel due to bandwidth limitation or congestion. Hence, it is mandatory to improvise the existing control strategies for NCS. In this chapter, we propose an approach for designing discrete-time sliding mode controller for NCS having random fractional delay and multiple packet loss simultaneously. The fractional delay that occurs within sampling period is modelled using Poisson's distribution function and is approximated using Thiran's delay approximation technique. The multiple packet loss that occurs in communication channel between sensor and controller is treated with uniform probability distribution function and compensated at controller end. Based on the proposed approach, a sliding surface is designed and is used to derive discrete-time sliding mode control law that computes the control actions in the presence of random network delay and multiple packet loss. Further, a second-order disturbance estimator is incorporated at the plant side to estimate the disturbance that occurs in the plant. The disturbance estimator guarantees the width of quasi-sliding mode band (QSMB) of order $O(h^3)$ with decreasing reaching steps. Hence, the robustness properties of closed-loop NCS are improved. The stability of the closed-loop NCS is also derived using Lyapunov approach that assures the finite-time state convergence in the presence of network non-idealities. The efficacy of the proposed algorithm is examined through simulation results.

Keywords Discrete-time sliding mode control · Networked Control System Disturbance estimation · Communication delay · Packet loss

7.1 Networked Control Systems

Figure 7.1 depicts the block diagram of Networked Control System (NCS) with multiple packet transmission policy and disturbance observer. As shown in the figure

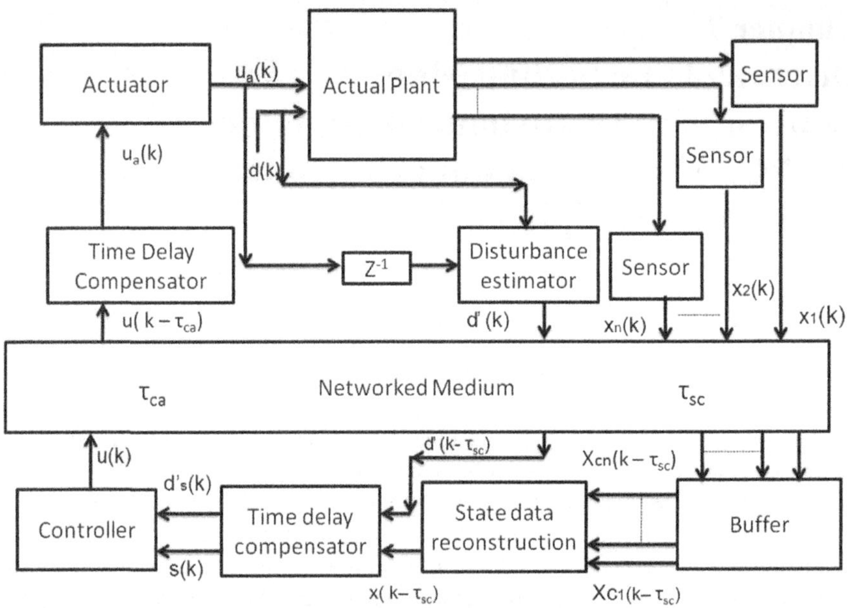

Fig. 7.1 Block diagram of NCS with multiple packet transmission

the plant, sensor and disturbance observer are connected to controller through the communication medium in the forward channel. And the controller output is also connected to plant actuator through communication channel. As shown the state information is measured from multiple sensors located at different places and is transmitted through the network in the form of packets. These multiple packets will suffer from random delay while transmitted from sensor to controller. In this multiple packet transmission policy, it is necessary to consider three different situations (i) none of the packets are lost, (ii) any one or more than one packet is lost or (iii) all the packets are lost [1–6]. If second or third situation arises during transmission, the frame structure would be incomplete at the controller side and the false control actions will be generated which may seriously deteriorate the performance of the closed-loop system.

Hence, we shall consider only multiple packet transmission resulting from the geographically dispersed sensors, and consequently, only the sensing data will be affected by multiple packet transmission.

7.2   Problem Formulation

Consider a continuous-time linear time-invariant SISO system with input delay (network delay) and matched uncertainty as:

$$\dot{x}(t) = Ax(t) + Bu(t - \tau) + Dd(t), \tag{7.1}$$

$$y(t) = Cx(t), \tag{7.2}$$

where $x \in R^n$ is system state vector, $u \in R^m$ is control input, $y \in R^p$ is system output, $A \in R^{n \times n}$, $B \in R^{n \times m}$, $C \in R^{p \times n}$, $D \in R^{n \times m}$ are the matrices of appropriate dimensions, $d(t)$ is the matched bounded disturbance with $|d(t)| \le d_{max}$, and τ is the total networked-induced delay in continuous-time domain.

The discrete-time system for the system (7.1) and (7.2) with sampling time h is obtained as:

$$x(k + 1) = Fx(k) + Gu(k - \hat{\tau}) + d(k), \tag{7.3}$$

$$y(k) = Cx(k), \tag{7.4}$$

where $F = e^{Ah}$, $G = \int_0^h e^{At} B dt$, $d(k) = \int_0^h e^{At} Dd((k + 1)h - t)dt \in O(h)$. Since $|d(t)| \le d_{max}$, it can be inferred that $d(k)$ is also bounded and $O(h)$ [7]. For simplicity, it is assumed that $d(k)$ is slowly varying and remains constant over the interval $kh \le t \le (k + 1)h$ [7].

In conventional NCS having single channel, the total network-induced fractional delay $(\hat{\tau}_t)$ is the combination of sensor to controller delay $(\hat{\tau}_{sc})$ and controller to actuator delay $(\hat{\tau}_{ca})$ which is given as,

$$\hat{\tau}_t = \hat{\tau}_{sc} + \hat{\tau}_{ca}, \tag{7.5}$$

where $\hat{\tau}_{sc} = \frac{\tau_{sc}}{h}$, $\hat{\tau}_{ca} = \frac{\tau_{ca}}{h}$, τ_{sc} is sensor to controller delay and τ_{ca} is controller to actuator delay.

However in the case of multiple sensor which is geographically distributed (refer Fig. 7.1), the total networked delay defined in Eq. (7.5) will not hold true as each sensor data will experience different network delays independent of one another. Thus, it is reasonable to assume max delay experienced by the packet as the sensor to controller delay $\hat{\tau}_{sc_{max}}$ under the multiple packet transmission policy which is given by:

$$\hat{\tau}_{sc_{max}} = max\{\hat{\tau}_{sc_i} : i = 1, 2, \ldots, n\}, \tag{7.6}$$

where $\hat{\tau}_{sc_i} = \frac{\tau_{sc_i}}{h}$ is the sensor to controller fractional delay generated from ith sensor.

Hence, the total network-induced fractional delay $(\hat{\tau})$ for multiple packet transmission is given as,

$$\hat{\tau} = \hat{\tau}_{sc_{max}} + \hat{\tau}_{ca}, \tag{7.7}$$

Further for analysis purpose, the total network-induced fractional delay $(\hat{\tau})$ occurring within the network is defined as,

$$\hat{\tau} = \frac{\tau}{h},$$

where h is the sampling interval and τ is the total networked-induced delay.

The following assumptions are made while deriving the discrete-time sliding mode control (DSMC) law:

Assumption 10 It is assumed that sensor to controller delay, controller to actuator delay and total network-induced delay are random and fractional in nature satisfying,

$$\hat{\tau}_{sc_i} \prec \hat{\tau} - \hat{\tau}_{ca}. \tag{7.8}$$

$$\hat{\tau}_{ca} \prec \hat{\tau}. \tag{7.9}$$

And,

$$\hat{\tau}_l \leq \hat{\tau} \leq \hat{\tau}_u, \tag{7.10}$$

where $\hat{\tau}_l$ and $\hat{\tau}_u$ indicate the lower and upper bounds of total network-induced delays.

Assumption 11 Here, it is assumed that multiple packet loss occurs from sensor to controller only. The assumption justifies that when time delay is more than sampling period the packet is considered as dropped or lost packet.

Assumption 12 The matched disturbance $d(t)$ is slowly time-varying and is bounded by upper and lower bounds as:

$$d_l \leq d(t) \leq d_u, \tag{7.11}$$

where d_l and d_u denote lower and upper bounds of disturbance.

Remark 18 It may be noted that in NCS apart from network-induced time delay there are also other delays like processing time delay of sensor signal ($\hat{\tau}_{sp}$), controller computational time delay ($\hat{\tau}_{cp}$) and actuator processing time delay ($\hat{\tau}_{ap}$) which has insignificant values as compared to communication delay and hence neglected.

Problem Statement: The main objective is to design the discrete-time sliding mode control (DSMC) law for system (7.3, 7.4) in the presence of random fractional delay $\hat{\tau}_{sc_i}$ and $\hat{\tau}_{ca}$ with multiple packet loss situation and matched uncertainty satisfying (7.11).

The next section discusses the mathematical policy of multiple packet loss that is generated from sensor to controller using the concept of uniform probability distribution function.

7.3 Multiple Packet Loss Policy

7.3.1 Multiple Packet Loss Policy from Sensor to Controller

As discussed earlier, the multiple packet loss situation arises where the sensors are geographically dispersed and each sensor measures the state information as mentioned in Eq. (7.3):

$$x(k) = \left[X_1^T(k) \ X_2^T(k) \ ... \ X_n^T(k) \right]^T , \tag{7.12}$$

where $[X_1(k), X_2(k),...,X_n(k)] \in R^1$.

To derive the multiple packet loss policy in forward channel, let us consider the case where n number of sensors located at different locations and far from each other are transmitting the state information through the communication channel as shown in Fig. (7.1).

The above state packet loss situation can be mathematically defined using probability distribution function as:

$$Xc_1(k) = \begin{cases} X_1(k - \hat{\tau}_{sc_1}); & \text{if } \rho_1 \prec P_{loss1} \\ X_1((k-1) - \hat{\tau}_{sc_1}); & \text{if otherwise} \end{cases}$$

$$Xc_2(k) = \begin{cases} X_2(k - \hat{\tau}_{sc_2}); & \text{if } \rho_2 \prec P_{loss2} \\ X_2((k-1) - \hat{\tau}_{sc_2}); & \text{if otherwise} \end{cases}$$

Similarly,

$$Xc_n(k) = \begin{cases} X_n(k - \hat{\tau}_{sc_i}); & \text{if } \rho_n \prec P_{lossn} \\ X_n((k-1) - \hat{\tau}_{sc_i}); & \text{if otherwise} \end{cases}$$

where $0 \prec P_{loss1} \prec 1, 0 \prec P_{loss2} \prec 1,..., 0 \prec P_{lossn} \prec 1$ are the probability of multiple state packet loss over the network and $\rho_1, \rho_2,...,\rho_n$ are the random variables uniformly distributed over the interval $[0,1]$.

Here, it may be understood that the packets transmitted by each sensor are received at controller side at different instances as the delay experienced by each packet is different. So all the state packets transmitted by each sensor arrived during each sampling period are buffered at controller side first, and then, the state vector is reconstructed at each sampling period considering the network delay $\hat{\tau}_{sc_i} \prec \hat{\tau}_{sc_{max}}$. The reconstructed state vector is used for controller design.

Thus, the communicated state variable over the network having the random delays from sensor to controller is given by:

$$x(k - \hat{\tau}_{sc_{max}}) = \begin{bmatrix} Xc_1(k - \hat{\tau}_{sc_1}) \\ Xc_2(k - \hat{\tau}_{sc_2}) \\ \cdot \\ \cdot \\ \cdot \\ \cdot \\ Xc_n(k - \hat{\tau}_{sc_i}) \end{bmatrix}.$$

Once the mathematical model of multiple packet loss is derived, the next step is to design the compensated sliding surface and discrete-time control law [8] that compensates the effect of random fractional delays in the presence of multiple packet loss and matched uncertainty.

7.4 Design of Sliding Surface with Multiple Packet Loss

In the literature, various time delay approximation techniques like Tustin approximation and bilinear transformation are widely used methods in discrete-time domain. But these methods are used for integer-type time delays only. Thiran's approximation technique [9] is widely used approximation technique for non-integer or fractional delay in signal processing application. Here, we are using the same method for compensating fractional time delay for designing the discrete-time sliding surface and sliding mode control.

The fractional delay is compensated using Thiran's approximation technique as under:

$$z^{-v} = \Sigma_{l=0}^{N}(-1)^l \binom{N}{l} \Pi_{j=0}^{N} \frac{2\hat{\tau}_{sc} + j}{2\hat{\tau}_{sc} + l + j} z^{-l}, \tag{7.13}$$

where N indicates the order of approximation, $v = \frac{\delta}{h}$ indicates the fractional delay in discrete domain, δ is the transmission delay of data packets, and h is the sampling interval.

The order of approximation is determined through *ceil* operator which rounds the nearest positive integer greater than or equal to v. It is given by:

$$N = ceil(v), \tag{7.14}$$

The designing of compensated sliding surface under multiple packet transmission with random fractional delay compensation is presented in the form of Lemma 5.

Lemma 5 *The compensated sliding surface for the given system* (7.3, 7.4) *with sensor to controller random fractional delay under multiple packet transmission and matched uncertainty satisfying condition* (7.10) *and* (7.11) *is given as:*

$$s(k) = C_s x(k) - \varsigma_{max} C_s x(k-1), \tag{7.15}$$

where C_s is the sliding gain and ς_{max} is the parameter designed using Thiran's approximation for max delay experienced by the packet as sensor to controller random fractional delay ($\hat{\tau}_{sc_{max}}$).

Proof In discrete-time domain, Poisson's distribution is the most suitable approach to modelled random fractional delay. This technique is used to represent the random variables of smaller values on the basis of number of events occurred over the specified interval of time [10]. The occurrence of the events is based on the number of trials required to generate the event. So, using this approach the random fractional delay is modelled by assuming that at each sampling instant an event takes place such that the networked delay might be lesser or higher than sampling interval.
The sliding variable with sensor to controller random fractional delay and multiple packet policy is given as:

$$s(k) = C_s x(k - \hat{\tau}_{sc_{max}}). \tag{7.16}$$

where $\hat{\tau}_{sc_{max}}$ is max random fractional delay from sensor to controller experienced by the packet in discrete-time domain.
The sensor to controller fractional delay $\{\hat{\tau}_{sc}\}$ for each sensor data is modelled using Poisson's distribution with probabilities given by,

$$Pr\{\hat{\tau}_{sc} = d_v\} = E\{d_v\} = \beta_v, \, v = 1, 2, \ldots, q. \tag{7.17}$$

where β_v is the positive scalar and $\Sigma_{v=0}^{q}\beta_v = 1$, $E\{d_v\}$ is the expectation of the stochastic variable d_v. The mathematical representation of β_v with Poisson's distribution is given by:

$$\beta_v = \frac{\lambda^w e^{-\lambda}}{w!}; \, w = 0, 1, 2, 3, \ldots \tag{7.18}$$

where w indicates the number of trials, λ denotes average number of events per interval, and e denotes the Euler's number.
Applying z-transform to above Eq. (7.16), we get

$$s(k) = C_s[x(z)z^{-\hat{\tau}_{scmax}}]. \tag{7.19}$$

Using Thiran's approximation and Eq. (7.13), $z^{-\hat{\tau}_{scmax}}$ is given as

$$s(k) = C_s x(z)[1 - \varsigma_{max}z^{-1}], \tag{7.20}$$

where $\varsigma_{max} = \frac{\hat{\tau}_{scmax}}{\hat{\tau}_{scmax}+1}$.
Applying inverse z-transform to above Eq. (7.20), we get

$$s(k) = C_s[x(k) - \varsigma_{max}x(k-1)]. \tag{7.21}$$

Further simplification,

$$s(k) = C_s x(k) - \varsigma_{max} C_s x(k-1). \tag{7.22}$$

This completes the *proof*.

The above Eq. (7.22) represents compensated sliding surface with multiple packet transmission policy. It is noticed that at each sampling interval the effect of delayed sensor packets is compensated through immediate past-state data packets, current-state data packets and parameter ς_{max}, while the effect of multiple packet loss is compensated through probability distribution function. If random variable ρ_n is greater than packet loss probabilities, then the data packet $Xc_n(k)$ is received, and if random variable ρ_n is lesser than packet loss probabilities, then $Xc_n(k-1)$ will be received at the controller.

  The next section represents the design of discrete-time sliding mode control law along with stability analysis in the presence of multiple packet transmission and matched uncertainty.

7.5   Discrete-Time Sliding Mode Control Law

This section discusses the derivation of discrete-time sliding mode control law for NCS using sliding surface (7.22) in the form of Theorem 7.1.

Theorem 7.1 *The discrete-time sliding mode control law for system (7.3, 7.4) in the presence of sensor to controller random fractional delay ($\hat{\tau}_{sc_i}$), multiple packet transmission and matched uncertainty $d(k)$ satisfying (7.11) is given as,*

$$u(k) = -(C_s G)^{-1}[Hx(k) - Ix(k) - J(s(k)) + \Delta_s(k)]. \tag{7.23}$$

where
$H = (C_s F), I = \varsigma_{max} C_s, J = \{1 - q[s(k)]\}$ *and* $\Delta_s(k) = 2d(k-1) - d(k-2) + \gamma d(k-1) - 2\gamma d(k-2) + \gamma d(k-3).$

Proof Let us reconstruct the reaching law in [11] in the presence of networked fractional delay, multiple packets transmission and disturbance estimator as:

$$s[(k+1)h] = \{1 - q[s(k)]\}s(k) + \hat{d}_s(k), \tag{7.24}$$

where $\hat{d}_s(k)$ is estimated disturbance applied to the controller, $q[s(k)] = \frac{\psi}{\psi + |s(k)|}$ with ψ as user-defined constant satisfying $\psi \geq d_2$ which is the deviated value of $d(k)$.

Remark 19 The estimated disturbance $\hat{d}(k)$ appearing in the reaching law is applied through the network. So, the compensated disturbance estimated signal according to Thiran's approximation [9] is computed as:

$$\hat{d}_s(k) = \hat{d}(k) - \gamma\hat{d}(k-1) \tag{7.25}$$

where $\gamma = \frac{\hat{\tau}_{sc}}{\hat{\tau}_{sc}+1}$, $\hat{d}(k)$ is the output of disturbance estimator. According to [12] and [13], the disturbance $d(k)$ possesses following *lemma*.

Lemma 6 $d(k) = O(h), d(k) - d(k-1) = O(h^2)$ *and* $d(k) = d(k) - 2d(k-1) + d(k-2) = O(h^3)$, *where h is the sampling time interval.*

Thus from above lemma 6, disturbance estimator designed is based on $O(h^3)$ since it provides better impact on the width of QSMB and increases the robustness properties of the designed controller. Thus, the output of disturbance estimator is given by:

$$\hat{d}(k) = d(k) - 2d(k-1) + d(k-2), \tag{7.26}$$

where disturbance $d(k-1)$ and $d(k-2)$ is estimated as follows:

$$d(k-1) = x(k) - Fx(k-1) - Gu_a(k-1), \tag{7.27}$$

where $u_a(k-1)$ is past compensated control signal available at the actuator side that is discussed in later part of this paper.

Remark 20 The output signal of disturbance estimator does not affect the states of the system since the matched uncertainty applied at the input of the system is slowly time-varying in nature.

The reaching law in Eq. (7.24) indicates that the system states always move towards the specified sliding band given as:

$$|s(k)| \leq O(h^3), \tag{7.28}$$

Substituting the value of $s(k+1)$ in Eq. (7.24), we get

$$C_s x(k+1) - \varsigma_{max}C_s x(k) = \{1 - q[s(k)]\}s(k) - \hat{d}_s(k), \tag{7.29}$$

Remark 21 It is noticed from Eq. (7.22) that random sensor to controller fractional delay is compensated at the sliding surface, while random controller to actuator fractional delay is compensated at the actuator side. So, the effect of delay will not be observed at the controller side for computing the control signal as well at actuator side. Thus without loss of generality, the control signal in Eq. (7.3) is given as

$$u(k - \tau') = u(k) \tag{7.30}$$

Substituting the value of $x(k+1)$,

$$C_s[Fx(k) + G(u(k) + d(k))] - \varsigma_{max} C_s x(k) = \{1 - q[s(k)]\}s(k) - \hat{d}_s(k). \quad (7.31)$$

Further simplification gives

$$C_s Fx(k) + C_s G(u(k)) + d(k) - \varsigma_{max} C_s x(k) = \{1 - q[s(k)]\}s(k) + \hat{d}(k) - \gamma\hat{d}(k). \quad (7.32)$$

Solving above Eq. (7.32) and using (7.26), the proposed discrete-time sliding mode control law can be expressed as:

$$u(k) = -(C_s G)^{-1}[Hx(k) - Ix(k) - J(s(k)) + \Delta_s(k)]. \quad (7.33)$$

where
$H = (C_s F)$, $I = \varsigma_{max} C_s$, $J = \{1 - q[s(k)]\}$ and $\Delta_s(k) = 2d(k-1) - d(k-2) + \gamma d(k-1) - 2\gamma d(k-2) + \gamma d(k-3)$.

This completes the *proof*.

It can be observed from (7.33) that the control law is formed by compensated state variables, sliding variable and compensated past disturbance signal multiplied with parameter γ. Thus, it is clear that due to disturbance estimator the controller does not require the knowledge of the bounds of disturbances applied at the input side of the channel. This makes the proposed controller more robust in nature.

The computed control action generated from (7.33) is applied to the actuator, will experience controller to actuator fractional delay. Thus using Thiran's approximation, the compensated control signal available at the actuator side is given as,

$$u_a(k) = u(k) - \varsigma' u(k-1), \quad (7.34)$$

$\varsigma' = \frac{\hat{\tau}_{ca}}{1+\hat{\tau}_{ca}}$.
From Eq. (7.34), it can be noticed that at each sampling instant the effect of random fractional delay from controller to actuator is compensated in control signal through the proposed technique.

The next section discusses about the stability of closed-loop system such that the system states remain within the specified band using control law (7.33).

7.6　Stability Analysis

Theorem 7.2 *The trajectories of the closed-loop system (7.3, 7.4) with the controller mentioned in Eq. (7.33) in the presence of random fractional delay ($\hat{\tau}$) satisfying (7.11), multiple packet loss and matched uncertainty $d(k)$ drive towards the sliding surface (7.22) such that the following condition holds true:*

$$\eta s^T(k)s(k) \succ 0. \tag{7.35}$$

Proof The compensated sliding surface in (7.22) is given by:

$$s(k) = C_s x(k) - \varsigma_{max} C_s x(k-1). \tag{7.36}$$

Let us consider Lyapunov function as,

$$V_s(k) = s^T(k)s(k). \tag{7.37}$$

Taking forward difference we have,

$$\Delta V_s(k) = s^T(k+1)s(k+1) - s^T(k)s(k). \tag{7.38}$$

Using Eq. (7.22), we get

$$\Delta V_s(k) = [C_s x(k+1) - \varsigma_{max} C_s x(k)]^T [C_s x(k) \\ -\varsigma_{max} C_s x(k-1)] - s^T(k)s(k). \tag{7.39}$$

Substituting the value of $x(k+1)$, we get

$$\Delta V_s(k) = [C_s[Fx(k) + G(u(k)) + d(k)] \\ -\varsigma_{max} C_s x(k)]^T [C_s[Fx(k) + G(u(k)) + d(k)] \\ -\varsigma_{max} C_s x(k-1)] - s^T(k)s(k). \tag{7.40}$$

Substituting the value of $u(k)$ and further simplifying it, we get

$$\Delta V_s(k) = \kappa - s^T(k)s(k). \tag{7.41}$$

where
$\kappa = [[1 - q(s(k))]s(k) + \Delta_s]^T [[1 - q(s(k))]s(k) + \Delta_s]$. The term κ can be tuned closed to zero by appropriately selecting the parameter ψ. Thus, it can be noticed that even though in stability analysis the knowledge of the bounds of disturbances is not needed due to disturbance rejection. So, if κ is closed to zero, then $s^T(k)s(k)$ will be larger than κ. Thus for any small parameter η, we have

$$\Delta V_s(k) \prec \eta s^T(k)s(k). \tag{7.42}$$

Thus, by tuning the parameter ψ, we have $\Delta V_s(k) \prec \eta s^T(k)s(k)$ which guarantees the convergence of $\Delta V_s(k)$ and implies that any trajectory of the system (7.3, 7.4) will be driven onto the sliding surface and maintained on it.
This completes the *proof*.

7.7 Results and Discussions

In this section, efficacy of designed control algorithm is validated in the presence of random fractional delay, multiple packet transmission with losses and matched uncertainty applied at the input channel of the system. The simulation results are carried out using illustrative example in [14]. Consider the continuous-time LTI system as,

$$\dot{x}(t) = Ax(t) + Bu(t - \tau) + Dd(t), \tag{7.43}$$

$$y(t) = Cx(t), \tag{7.44}$$

where

$$A = \begin{bmatrix} -0.7 & 2 \\ 0 & -1.5 \end{bmatrix}, \ B = \begin{bmatrix} -0.03 \\ -1 \end{bmatrix},$$

$$C = \begin{bmatrix} 1 & 0 \end{bmatrix}, \ D = \begin{bmatrix} 1 \\ 1 \end{bmatrix}, \ d(t) = 0.2 sin(0.086t).$$

Discretizing the above system with sampling interval of $h = 30$ ms, we get

$$x(k + 1) = Fx(k) + Gu(k - \hat{\tau}) + d(k), \tag{7.45}$$

$$y(k) = Cx(k), \tag{7.46}$$

where

$$F = \begin{bmatrix} 0.9792 & 0.05805 \\ 0 & 0.956 \end{bmatrix}, \ G = \begin{bmatrix} -0.001771 \\ -0.02934 \end{bmatrix},$$

$$C = \begin{bmatrix} 1 & 0 \end{bmatrix}.$$

The plant consists of two sensor data signals $x(k) = [X_1^T(k) \ X_2^T(k)]^T$ that are located at far distances from one another. Figures 7.2 and 7.3 show the nature of compensated and delayed state packets, sliding surface, control signal computed at controller side, compensated control signal computed at actuator side and disturbance estimator under multiple packet transmission. Figures 7.4a–f and 7.5a, b show the nature of compensated control signal at the actuator side and delayed control signal under multiple packet loss situations and random fractional delays greater than sampling interval. Figure 7.5c–e shows the response of network-induced delay modelled using Poisson's distribution. It is assumed that the probability of total networked delay lesser than sampling interval is $p = 0.75$ while the probability of total networked delay greater than sampling interval is $p = 0.25$. This assumption indicates that according to Poisson's distribution at every sampling interval at least one event is generated with zero trial. Thus, the total network delay range computed through Poisson's distribution is 0.0003 s $\leq \tau \leq 0.042$ s, respectively (see Fig. 7.5e). So, based on above distribution sensor to controller delay for both sensors is 0.2 ms $\leq \tau_{sc_1}, \tau_{sc_2} \leq 12$ ms (see Fig. 7.5c) and controller to actuator delay is 0.1 ms $\leq \tau_{ca} \leq 30$ ms (see Fig. 7.5d) satisfying condition (7.10) and (7.11), respectively. In order to prove the robustness of the proposed control algorithm slowly

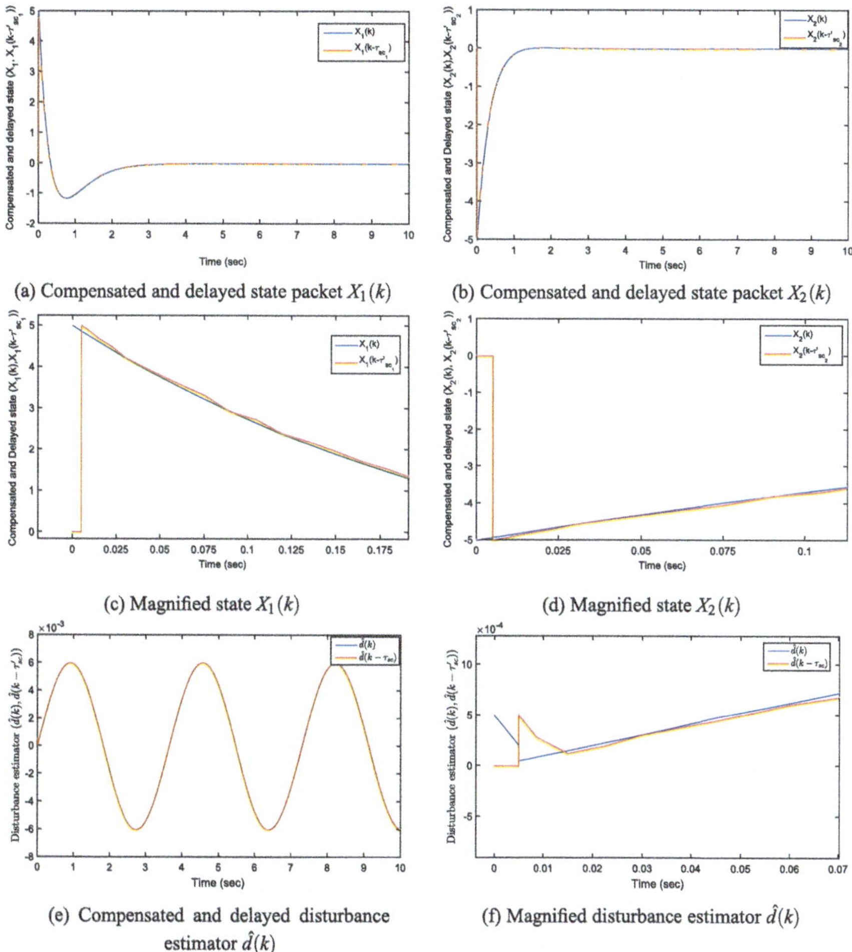

(a) Compensated and delayed state packet $X_1(k)$

(b) Compensated and delayed state packet $X_2(k)$

(c) Magnified state $X_1(k)$

(d) Magnified state $X_2(k)$

(e) Compensated and delayed disturbance estimator $\hat{d}(k)$

(f) Magnified disturbance estimator $\hat{d}(k)$

Fig. 7.2 **a–d** Actual and magnified simulated results of compensated and delayed state variables and **e–f** actual and magnified simulated results of compensated and delayed disturbance estimator

time-varying disturbance is applied to the input of the system. Figure 7.5f shows the result of actual disturbance and estimated disturbance at the plant side. It is observed that the disturbance estimator designed using $O(h^3)$ estimates incoming disturbance very precisely at each sampling interval.

The sliding gain parameter computed using discrete LQR method with $Q = diag(1000, 1000)$ and $R = 1$ is $C_s = [-1.4025 \ -2.6345]$ having $\psi = 100$. Thus using (7.28) and referring Fig. 7.5f, the quasi-sliding mode band comes out to be $|s(k)| \preceq +0.006$ to -0.006, respectively.

As mentioned in Fig. 7.1, the state variables and disturbance estimator signals are transmitted through networks from plant side to controller side, while at actuator

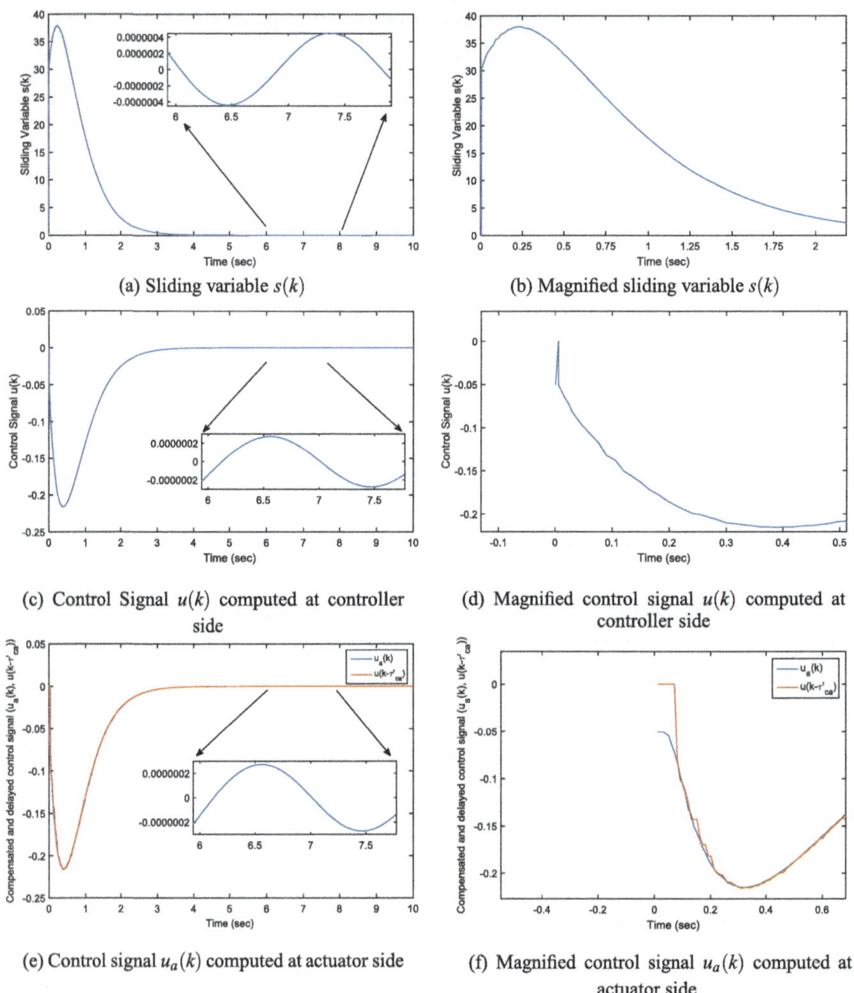

(a) Sliding variable $s(k)$

(b) Magnified sliding variable $s(k)$

(c) Control Signal $u(k)$ computed at controller side

(d) Magnified control signal $u(k)$ computed at controller side

(e) Control signal $u_a(k)$ computed at actuator side

(f) Magnified control signal $u_a(k)$ computed at actuator side

Fig. 7.3 **a–b** Actual and magnified simulated results of sliding variable, **c–d** actual and magnified simulated results of control signal at controller side and **e–f** actual and magnified simulated results of compensated control signal at actuator side and delayed control signal

side control signal is received through the networks. So, the time delay compensation algorithm should work efficiently in state variables, disturbance estimator and control signal. In order to show the efficacy of proposed compensation algorithm, each parameter is compared with its delayed parameter that is available after transmission through the networks.

Figure 7.2a, b shows the nature of compensated and delayed state packets with initial conditions $[5 \ -5]$, respectively. It is observed that the compensated and delayed state packets both converge to zero from their specified initial condition in the presence of specified random fractional delay and multiple packet transmission. In order

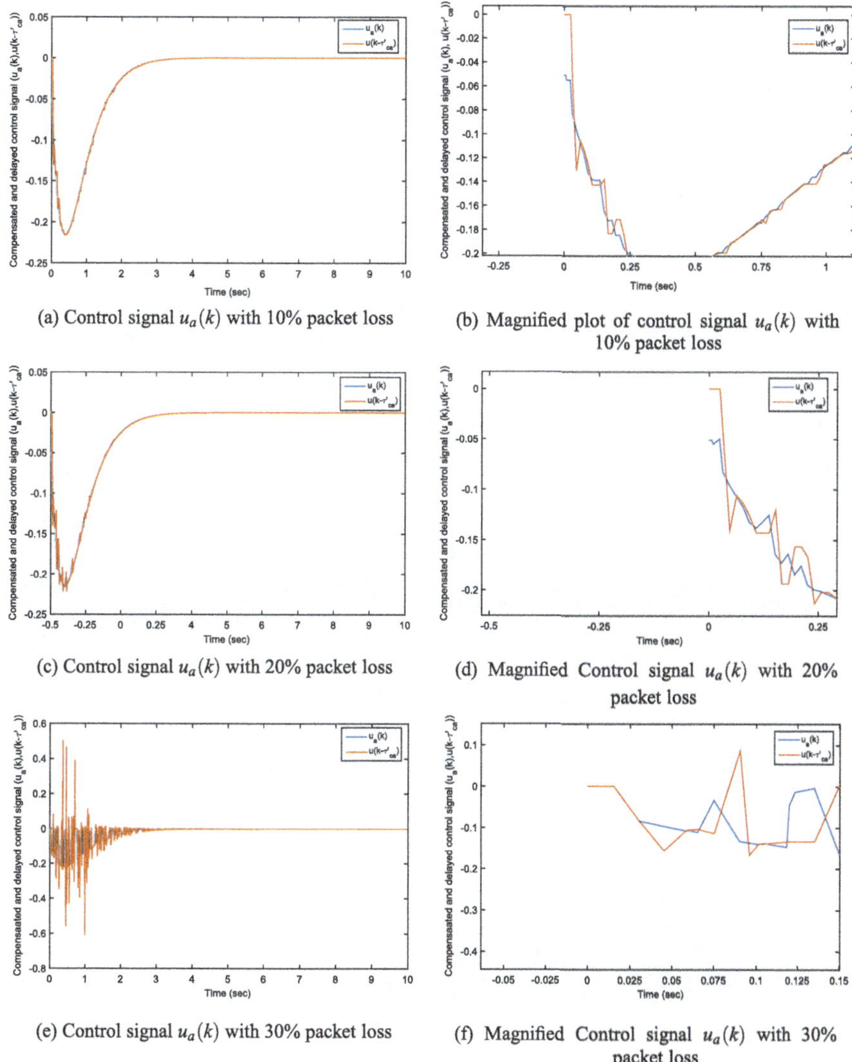

(a) Control signal $u_a(k)$ with 10% packet loss

(b) Magnified plot of control signal $u_a(k)$ with 10% packet loss

(c) Control signal $u_a(k)$ with 20% packet loss

(d) Magnified Control signal $u_a(k)$ with 20% packet loss

(e) Control signal $u_a(k)$ with 30% packet loss

(f) Magnified Control signal $u_a(k)$ with 30% packet loss

Fig. 7.4 a–f Actual and magnified simulated results of control signal at actuator side and delayed control signal with 10, 20 and 30% packet loss

to determine the exact effect of random fractional delay compensation, the magnified results of the same are shown in Fig. 7.2c, d, respectively. It is noticed that the compensated state packets in both cases are computed from first sampling instant, while the state packets available after the network are computed after the forward channel delay. Moreover, it is also observed that compensated state packets have better performance than delayed state packets. Figure 7.2e, f shows the nature of delayed and compensated disturbance estimator computed after the network. It can be noticed

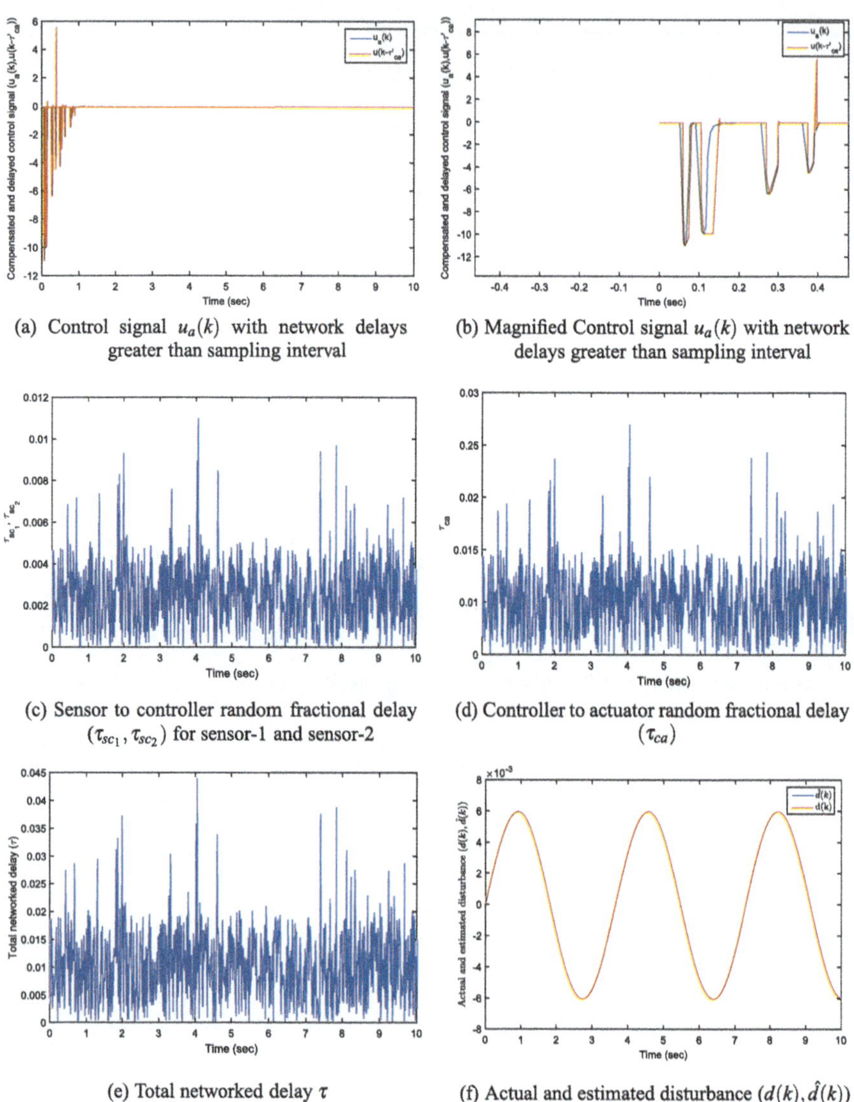

(a) Control signal $u_a(k)$ with network delays greater than sampling interval

(b) Magnified Control signal $u_a(k)$ with network delays greater than sampling interval

(c) Sensor to controller random fractional delay $(\tau_{sc_1}, \tau_{sc_2})$ for sensor-1 and sensor-2

(d) Controller to actuator random fractional delay (τ_{ca})

(e) Total networked delay τ

(f) Actual and estimated disturbance $(d(k), \hat{d}(k))$

Fig. 7.5 **a–b** Actual and magnified simulated results of control signal at actuator side with random fractional delays greater than sampling interval, **c–e** simulated results networked delay using Poisson's distribution and **f** simulated results of actual and estimated disturbance

from magnified result Fig. 7.2f that the effect of sensor to controller delay is also compensated at the controller side through time delay compensator. Figure 7.3a, b shows the nature of sliding variable and control signal computed at controller side through compensated state information available from state reconstruction block and compensated disturbance estimator signal available from time delay compensator. Observing the results, it is noticed that the width of QSMB (mentioned in small window of Fig. 7.3a is much more smaller than order $O(h^3)$ that causes the sliding surface to slide onto the origin having less chattering and decreasing reaching time in the presence of multiple packet transmission and network delay. The same effect of robustness is observed in control signal computed at controller side (see Fig. 7.3c) and compensated control signal computed at actuator side (see Fig. 7.3e). Thus, it proves that the compensated disturbance signal applied from the disturbance estimator of order $O(h^3)$ to the controller makes the proposed control algorithm more robust even in the presence of network non-idealities without prior knowledge of the bounds of disturbance applied at input side of the channel. The magnified results of sliding variable and control signal computed at controller side shown in Fig. 7.3b, d also justify that the effect of sensor to controller random fractional delay is compensated in the forward channel. Thus, it can be extended that the proposed discrete-time sliding mode controller designed in (7.32) is more robust in nature as it has ability to compensate the random fractional delay and reject the unknown disturbances applied at the input side of the channel.

Once the control actions are computed at the controller side, they are transmitted to actuator through the networks. Figure 7.3e shows the result of compensated control signal computed at the actuator end and delayed control signal transmitted through the networks. Observing the magnified result in Fig. 7.3f, it is noticed that the effect of random fractional delay from controller to actuator is compensated in the control signal that is transmitted through the networks with the same robustness property.

The robustness of the proposed algorithm is further examined for multiple packet loss. It is assumed that more than one packet is lost in the forward channel during transmission. Figure 7.4a, c, e shows the nature of compensated control signal computed at actuator side and delayed control signal under 10, 20 and 30% multiple packet loss, respectively. It is observed that the effect of random fractional delay from controller to actuator is compensated precisely and robustness is also achieved in the case of 10 and 20% packet loss. While in 30% packet loss even though the robustness is achieved, the effect of random fractional delay is not compensated. The magnified results of the same are shown in Fig. 7.4b, d, f, respectively. The proposed control algorithm is further examined for the network delays greater than sampling interval. It can be noticed from Fig. 7.5a, b that when the probabilities for modelling the fractional delays are reversed than previous case, the robustness of the proposed technique becomes poor and cannot able to compensate the effect of network delay in the presence of multiple packet transmission and matched uncertainty.

Thus, from above results it is perceived that the proposed controller compensates the effect of specified random fractional delay with multiple packet loss and shows the stable response satisfying condition (7.41) in the presence of random fractional delay, multiple packet transmission and matched uncertainty. Moreover, the designed control algorithm also proves to be more robust as the width of QSMB is much more smaller than (7.28) and decreases the reaching steps at the same time in the presence of network non-idealities (random fractional delay, multiple packet loss and matched uncertainty).

7.8 Conclusion

In this chapter, time delay approximation algorithm is proposed in discrete-time domain that compensates the effect of variable fractional delay in the presence of multiple packet loss condition. The random fractional delay is modelled using Poisson's distribution, while multiple packet loss is modelled using uniform probability distribution function. The multiple packet loss model is derived for geographically located sensors. The effects of random fractional delay are compensated in forward as well as feedback channels using Thiran's approximation, while the robustness of the proposed controller is proven through disturbance estimator of order $O(h^3)$ that does not require knowledge of the bounds of unknown disturbances applied at input side of the channel. The effect of multiple packet loss is compensated in the forward channel such that if the packet is lost within the sampling interval, it will be recovered for that instant. The discrete-time sliding mode controller is designed based on the proposed sliding surface. The stability condition of the closed-loop system is derived using Lyapunov approach such that the system states remain within the specified band over a finite interval of time. The efficacy of the proposed control algorithm is checked under different conditions through illustrative example. The simulation results show that the discrete-time control law derived using Thiran's approximation technique compensates the effect of random fractional delay and increases robustness properties even in the presence of multiple packet loss with probability of 20% and matched uncertainty. In future, the same control algorithm can be extended for wireless NCS (WNCS).

References

1. D. Shah, A. Mehta, Multirate output feedback based discrete-time sliding mode control for fractional delay compensation in NCSs, in *IEEE Conference on Industrial Technology (ICIT-2017)* (2017), pp. 848–853
2. D. Shah, A. Mehta *"Discrete-Time Sliding Mode Control Using Thiran's Delay Approximation for Networked Control System"*, 43rd Annual Conference on Industrial Electronics (IECON-17), pp. 3025–3031, Nov. 2017
3. D. Shah, A.J. Mehta, Design of robust controller for networked control system, in *Proceedings of IEEE International Conference on Computer, Communication and Control Technology*, Sept. 2014, pp. 385–390
4. D. Shah, A. Mehta, Output feedback discrete-time networked sliding mode control, in *IEEE Proceedings of Recent Advances in Sliding Modes (RASM)* (2015), pp. 1–7

5. D. Shah, A. Mehta, Discrete-time sliding mode controller subject to real-time fractional delays and packet losses for networked control system. Int. J. Control Autom. Syst. (IJCAS) **15**(6), 2690–2703 (Dec. 2017)
6. D. Shah, A. Mehta, Fractional delay compensated discrete-time SMC for networked control system. Digit. Commun. Netw. (DCN), Elsevier, **2**(3), 385–390, Dec. 2016
7. A.J. Mehta, B. Bandyopadhyay, Multirate output feedback based stochastic sliding mode control. J. Dyn. Syst. Measur. Control **138**(12), 124503(1-6) (2016)
8. A.J. Mehta, B. Bandyopadhyay, A. Inoue, Reduced-order observer design for servo system using duality to discrete-time sliding-surface design. IEEE Trans. Ind. Electron. **57**(11), 3793–3800 (2010)
9. J. Thiran, Recursive digital filters with maximally flat group delay. IEEE Trans. Circ. Theory **18**(6), 659–664 (1971)
10. L. Brown, L. Zhao, A test of Poisson's Distribution. J. Stat. **64**(3), 611–625 (2002)
11. A. Bartoszewicz, P. Lesniewski, Reaching law approach to the sliding mode control of periodic review inventory systems. IEEE Trans. Autom. Sci. Eng. **11**(3), 810–817 (2014)
12. A. Bartoszewicz, Remarks on discrete-time variable structure control systems. IEEE Trans. Ind. Electron. **43**(1), 235–238 (1996)
13. A. Bartoszewicz, Discrete-time variable structure control systems. IEEE Trans. Ind. Electron. **45**(4), 633–637 (1996)
14. J. Wu, T. Chen, Design of networked control systems with packet dropouts. IEEE Trans. Autom. Control **52**(7), 1314–1319 (2007)

Chapter 8
Conclusion, Future Scope and Challenges

8.1 Conclusion and Future Scope

In this book, a novel method of designing discrete-time sliding mode controller in the presence of fractional delay along with packet loss occurrence due to network medium is presented. Firstly, the effect of deterministic type fractional delay is compensated using Thiran's approximation technique and the compensated state information is used for the design of sliding surface. A switching type discrete-time networked sliding mode controller is designed using the compensated sliding surface which computes the control sequences in the presence of matched uncertainty. The condition for stability of the closed-loop system is derived using Lyapunov approach. The efficacy of the proposed algorithms is tested on DC servo motor setup with different network delays and external disturbances. The results are compared with the conventional SMC, and it is observed that the DSMC design using Thiran's approximation compensates the fractional delays satisfactorily. The major drawback of switching type DSMC is that it induces unmodeled dynamics and the large QSMC band which compromise the robustness property. To overcome this issue, a non-switching type discrete-time sliding mode controller with Thiran's approximation technique is proposed. The efficacy of the proposed algorithm is tested through illustrative example as well as DC servo motor setup with different deterministic fractional delays and matched uncertainty. The results are also compared with switching type SMC as well as conventional SMC without delay compensation. It is observed that the proposed non-switching type DSMC algorithm not only rapidly converges but also reduces the amplitude of control signal and offers better fractional delay compensation. The efficacy of the proposed control algorithm is also tested under real-time networks like CAN and Switched Ethernet as a network medium in the presence of packet loss condition. From the simulation and experimental results, it is observed that the proposed DSMC law using Thiran's approximation compensates the effect of fractional network delay even in the presence of real-time networks and packet loss situation. The DSMC algorithms are further extended for output feedback control using the well-known multirate output feedback approach.

© Springer Nature Singapore Pte Ltd. 2018

D. H. Shah and A. Mehta, *Discrete-Time Sliding Mode Control for Networked Control System*, Studies in Systems, Decision and Control 132, https://doi.org/10.1007/978-981-10-7536-0_8

The DSMC control law is further extended for random fractional delays with single packet loss situation. The random fractional delay is modeled using Poisson's distribution function, and packet loss is modeled using probability distribution function. The stability condition derived for the closed-loop system ensures finite-time convergence in the presence of random fractional delay, packet loss and matched uncertainty. The effectiveness of the proposed control algorithm is examined through DC servo motor setup under random fractional delay and single packet loss. The results show that the DSMC control law derived using Thiran's approximation compensates random fractional delay accurately even in the presence of single packet loss with probability of 30% as well as networked delays having values greater than sampling interval.

Lastly, the DSMC algorithm is extended for random fractional delay with multiple packet loss situations. The multiple packet loss situation is modeled using probability distribution function, while random fractional delay is modeled using Poisson's distribution. The proposed DSMC control law effectively compensates the random fractional delay with multiple packet loss. The stability of the closed-loop NCS is assured through Lyapunov approach under multiple packet transmission. The efficacy of proposed DSMC algorithm is verified through DC servo motor plant under random fractional delay, multiple packet loss and matched uncertainty. The results show that the proposed control law compensates the effect of random fractional delay even in the presence of multiple packet loss with probability of 30% as well as networked delays greater than sampling interval.

In future, the DSMC algorithm can be extended for Wireless Networked Control System (WNCS). Also, the efficacy of the proposed algorithm needs to be checked for direct structure NCS, hierarchical structure and shared network structure NCS.

8.2 Challenges

In Networked Control System although much work has been done in last two decades. However, still there are various challenges that need to be solved while designing the control algorithm such as

- There are no standard algorithm available in the literatures that discusses the variable packet loss model. The main disadvantages in existing models are random time delay and packet loss lumped together. So it is difficult to differentiate the effect of losses in the system response. Thus, there is need for developing some model that takes care of only packet loss.
- In NCS, packet disorder is also one of the challenging issues that occurs in event-triggered model. In the literature, it is assumed that packet disorder does not take place in such type of models. However, in real-time applications this assumption does not hold true. So, there is a need of control algorithm that discusses the packet disorder in event-triggered model.

- The concept of distributed networked control is still an attractive and challenging area for the researchers in Networked Control System. In distributed networked control, large number of sensors and actuators are connected to the various controllers through different topologies which causes random time delay and packet loss at each level depending on the type of topology and communication medium. These factors make the analysis and synthesis more complicated and challenging for distributed networked control systems.
- There is also need to study the reliability issues for the Networked Control System.
- Investigations related to packet dropout and network-induced delays in industrial-based NCS are popular. However, studies related to the possible positive effects of these parameters are still not conducted. Designing of the controller which emphasizes these effects is interesting.
- Apart from random time delay and packet loss issues, there are various concerns such as bandwidth sharing, security and resource allocation in NCS that need to be studied for better performance of Networked Control System.

The manufacturer's authorised representative in the EU is Springer
Nature Customer Service Centre GmbH, Europaplatz 3, 69115 Heidelberg,
Germany. If you have any concerns regarding our products, please
contact ProductSafety@springernature.com

Printed and bound by CPI Group (UK) Ltd, Croydon, CR0 4YY
28/11/2025
02007692-0016